Principles and Techniques of Plant Breeding

Principles and Techniques of Plant Breeding

Edited by
Damien Miller

Larsen & Keller
www.larsen-keller.com

Principles and Techniques of Plant Breeding
Edited by Damien Miller
ISBN: 978-1-63549-222-4 (Hardback)

Larsen & Keller

Published by Larsen and Keller Education,
5 Penn Plaza,
19th Floor,
New York, NY 10001, USA

Cataloging-in-Publication Data

Principles and techniques of plant breeding / edited by Damien Miller.
p. cm.
Includes bibliographical references and index.
ISBN 978-1-63549-222-4
1. Plant breeding. 2. Plant breeding--Technique.
I. Miller, Damien.

SB123 .P75 2017
631.53--dc23

Printed and bound in China.

For more information regarding Larsen and Keller Education and its products, please visit the publisher's website www.larsen-keller.com

Table of Contents

Preface

This book aims to shed light on some of the unexplored aspects of plant breeding. Plant breeding is the method of enhancing favorable characteristics by artificially causing plants to breed. It uses a rigorous process where plants are selected and evaluated for effective outputs in the most scientific and systematic way. Selected concepts that re-design plant breeding have been presented in this book. The topics included in this text are of utmost significance and are bound to provide incredible insights to the readers. It aims to serve as a resource guide for graduates and postgraduates alike and contribute to the growth of the discipline.

To facilitate a deeper understanding of the contents of this book a short introduction of every chapter is written below:

Chapter 1- Plant breeding uses deliberate interbreeding of plants to produce new crop varieties or lines with desirable properties. It can be achieved by many techniques; some of these techniques are simple in form while others are more complex. The chapter on plant breeding offers an insightful focus, keeping in mind the complex subject matter.

Chapter 2- The chapter serves as a source to understand the methods in plant breeding. Reproduction, pollination, plant propagation and apomixis are examined in this chapter, for a better comprehension on plant breeding. The aspects elucidated in the following content are of vital importance, and provide a better understanding on plant breeding.

Chapter 3- Processes like molecular breeding, marker-assisted selection, doubled haploidy, genetic engineering, family based QTL mapping etc. are described in the following chapter in a critical and systematic manner. The following chapter elucidates the various processes and technologies that are related to plant breeding.

Chapter 4- Plant breeding comes with issues of concern, particularly in regard to food crop. The aim of this chapter is to provide an in-depth understanding on the issues and concerns of plant breeding. Nutrition and plant breeder rights are some of the significant and important topics related to plant breeding. The following chapter unfolds its crucial aspects in a critical yet systematic manner.

Chapter 5- The associated fields of plant breeding dealt within this chapter are cultivated plant taxonomy and organic farming. The study of the theory and practice of the science that identifies, describes and classifies those plants whose origin is primarily due to intentional human activity is known as cultivated plant taxonomy. The topics discussed in the chapter are of great importance to broaden the existing knowledge on plant breeding.

Chapter 6- Plant breeding started with the domestication of the first agricultural plants, almost 9000 to 11000 years back. Laws of inheritance formed the basis of the new science of genetics, causing many plant scientists to also research and to contribute to improving crop production through plant breeding. The evolution of plant breeding explained in this chapter is of great importance to broaden the existing knowledge on plant breeding.

Chapter 7- This chapter focuses on the acts affiliated to plant breeding. This chapter lists the following acts like the protection of plant varieties and farmers rights act, 2001, plant variety protection act of 1970, plant patent act of 1930 and international union for the protection of new varieties of plants. This chapter will help the readers in understanding these acts.

I would like to share the credit of this book with my editorial team who worked tirelessly on this book. I owe the completion of this book to the never-ending support of my family, who supported me throughout the project.

Editor

1

Introduction to Plant Breeding

Plant breeding uses deliberate interbreeding of plants to produce new crop varieties or lines with desirable properties. It can be achieved by many techniques; some of these techniques are simple in form while others are more complex. The chapter on plant breeding offers an insightful focus, keeping in mind the complex subject matter.

Plant breeding is the art and science of changing the traits of plants in order to produce desired characteristics. Plant breeding can be accomplished through many different techniques ranging from simply selecting plants with desirable characteristics for propagation, to more complex molecular techniques.

The Yecoro wheat (right) cultivar is sensitive to salinity, plants resulting from a hybrid cross with cultivar W4910 (left) show greater tolerance to high salinity

Plant breeding has been practiced for thousands of years, since near the beginning of human civilization. It is practiced worldwide by individuals such as gardeners and farmers, or by professional plant breeders employed by organizations such as government institutions, universities, crop-specific industry associations or research centers.

International development nation agencies believe that breeding new crops is important for ensuring food security by developing new varieties that are higher-yielding, disease resistant, drought-resistant or regionally adapted to different environments and growing conditions.

History

Plant breeding started with sedentary agriculture and particularly the domestication of the first agricultural plants, a practice which is estimated to date back 9,000 to 11,000 years. Initially early farmers simply selected food plants with particular desirable characteristics, and employed these as progenitors for subsequent generations, resulting in an accumulation of valuable traits over time.

Gregor Mendel's experiments with plant hybridization led to his establishing laws of inheritance. Once this work became well known, it formed the basis of the new science of genetics, which stimulated research by many plant scientists dedicated to improving crop production through plant breeding.

Modern plant breeding is applied genetics, but its scientific basis is broader, covering molecular biology, cytology, systematics, physiology, pathology, entomology, chemistry, and statistics (biometrics). It has also developed its own technology.

Classical Plant Breeding

One major technique of plant breeding is selection, the process of selectively propagating plants with desirable characteristics and eliminating or "culling" those with less desirable characteristics.

Another technique is the deliberate interbreeding (crossing) of closely or distantly related individuals to produce new crop varieties or lines with desirable properties. Plants are crossbred to introduce traits/genes from one variety or line into a new genetic background. For example, a mildew-resistant pea may be crossed with a high-yielding but susceptible pea, the goal of the cross being to introduce mildew resistance without losing the high-yield characteristics. Progeny from the cross would then be crossed with the high-yielding parent to ensure that the progeny were most like the high-yielding parent, (backcrossing). The progeny from that cross would then be tested for yield (selection, as described above) and mildew resistance and high-yielding resistant plants would be further developed. Plants may also be crossed with themselves to produce inbred varieties for breeding. Pollinators may be excluded through the use of pollination bags.

Classical breeding relies largely on homologous recombination between chromosomes to generate genetic diversity. The classical plant breeder may also make use of a num-ber of *in vitro* techniques such as protoplast fusion, embryo rescue or mutagenesis to generate diversity and produce hybrid plants that would not exist in nature.

Traits that breeders have tried to incorporate into crop plants include:

1. Improved quality, such as increased nutrition, improved flavor, or greater beauty
2. Increased yield of the crop
3. Increased tolerance of environmental pressures (salinity, extreme temperature, drought)

4. Resistance to viruses, fungi and bacteria
5. Increased tolerance to insect pests
6. Increased tolerance of herbicides
7. Longer storage period for the harvested crop

Before World War II

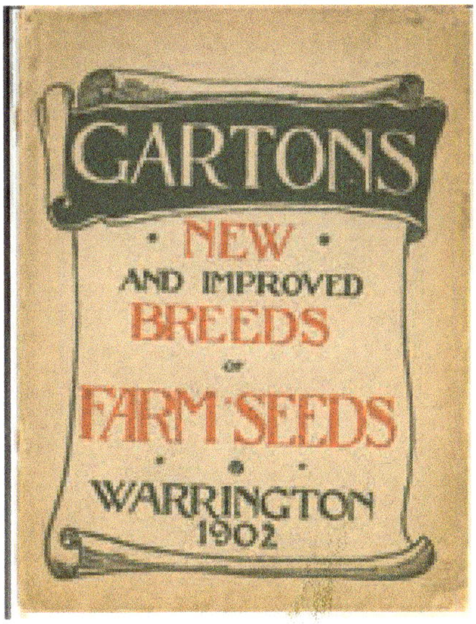

Garton's catalogue from 1902

Successful commercial plant breeding concerns were founded from the late 19th century. Gartons Agricultural Plant Breeders in England was established in the 1890s by John Garton, who was one of the first to commercialize new varieties of agricultural crops created through cross-pollination. The firm's first introduction was Abundance Oat, one of the first agricultural grain varieties bred from a *controlled* cross, introduced to commerce in 1892.

In the early 20th century, plant breeders realized that Mendel's findings on the non-random nature of inheritance could be applied to seedling populations produced through deliberate pollinations to predict the frequencies of different types. Wheat hybrids were bred to increase the crop production of Italy during the so-called "Battle for Grain" (1925–1940). Heterosis was explained by George Harrison Shull. It describes the tendency of the progeny of a specific cross to outperform both parents. The detection of the usefulness of heterosis for plant breeding has led to the development of inbred lines that reveal a heterotic yield advantage when they are crossed. Maize was the first species where heterosis was widely used to produce hybrids.

Statistical methods were also developed to analyze gene action and distinguish heritable variation from variation caused by environment. In 1933 another important breeding technique, cytoplasmic male sterility (CMS), developed in maize, was described by Marcus Morton Rhoades. CMS is a maternally inherited trait that makes the plant produce sterile pollen. This enables the production of hybrids without the need for labor-intensive detasseling.

These early breeding techniques resulted in large yield increase in the United States in the early 20th century. Similar yield increases were not produced elsewhere until after World War II, the Green Revolution increased crop production in the developing world in the 1960s.

After World War II

In vitro-culture of Vitis (grapevine), Geisenheim Grape Breeding Institute

Following World War II a number of techniques were developed that allowed plant breeders to hybridize distantly related species, and artificially induce genetic diversity.

When distantly related species are crossed, plant breeders make use of a number of plant tissue culture techniques to produce progeny from otherwise fruitless mating. Interspecific and intergeneric hybrids are produced from a cross of related species or genera that do not normally sexually reproduce with each other. These crosses are referred to as *Wide crosses*. For example, the cereal triticale is a wheat and rye hybrid. The cells in the plants derived from the first generation created from the cross contained an uneven number of chromosomes and as result was sterile. The cell division inhibitor colchicine was used to double the number of chromosomes in the cell and thus allow the production of a fertile line.

Failure to produce a hybrid may be due to pre- or post-fertilization incompatibility. If fertilization is possible between two species or genera, the hybrid embryo may abort before maturation. If this does occur the embryo resulting from an interspecific or intergeneric cross can sometimes be rescued and cultured to produce a whole plant. Such a method is referred to as Embryo Rescue. This technique has been used to produce new rice for Africa, an interspecific cross of Asian rice *(Oryza sativa)* and African rice *(Oryza glaberrima).*

Hybrids may also be produced by a technique called protoplast fusion. In this case protoplasts are fused, usually in an electric field. Viable recombinants can be regenerated in culture.

Chemical mutagens like EMS and DMS, radiation and transposons are used to generate mutants with desirable traits to be bred with other cultivars - a process known as *Mutation Breeding*. Classical plant breeders also generate genetic diversity within a species by exploiting a process called somaclonal variation, which occurs in plants produced from tissue culture, particularly plants derived from callus. Induced polyploidy, and the addition or removal of chromosomes using a technique called chromosome engineering may also be used.

When a desirable trait has been bred into a species, a number of crosses to the favored parent are made to make the new plant as similar to the favored parent as possible. Returning to the example of the mildew resistant pea being crossed with a high-yielding but susceptible pea, to make the mildew resistant progeny of the cross most like the high-yielding parent, the progeny will be crossed back to that parent for several generations. This process removes most of the ge-netic contribution of the mildew resistant parent. Classical breeding is therefore a cyclical process.

With classical breeding techniques, the breeder does not know exactly what genes have been introduced to the new cultivars. Some scientists therefore argue that plants produced by classical breeding methods should undergo the same safety testing regime as genetically modified plants. There have been instances where plants bred using classical techniques have been unsuitable for human consumption, for example the poison solanine was unintentionally increased to unacceptable levels in certain varieties of potato through plant breeding. New potato varieties are often screened for solanine levels before reaching the marketplace.

Modern Plant Breeding

Modern plant breeding may use techniques of molecular biology to select, or in the case of genetic modification, to insert, desirable traits into plants. Application of biotechnology or molecular biology is also known as molecular breeding.

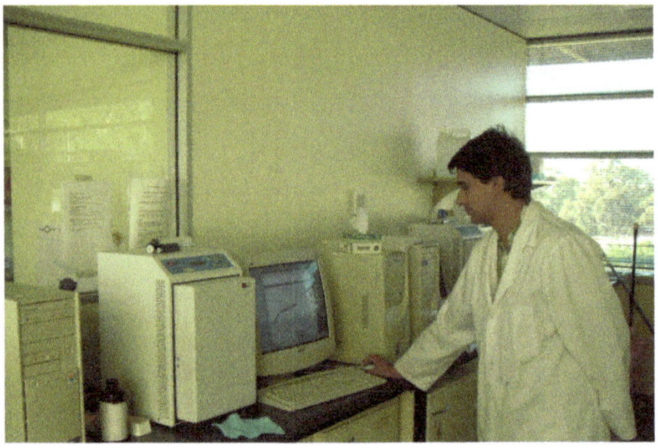

Modern facilities in molecular biology have converted classical plant breeding to molecular plant breeding

Steps of Plant Breeding

The following are the major activities of plant breeding:

1. Collection of variation
2. Selection
3. Evaluation
4. Release
5. Multiplication
6. Distribution of the new variety

Marker Assisted Selection

Sometimes many different genes can influence a desirable trait in plant breeding. The use of tools such as molecular markers or DNA fingerprinting can map thousands of genes. This allows plant breeders to screen large populations of plants for those that possess the trait of interest. The screening is based on the presence or absence of a certain gene as determined by laboratory procedures, rather than on the visual identification of the expressed trait in the plant.

Reverse Breeding and Doubled Haploids (DH)

A method for efficiently producing homozygous plants from a heterozygous starting plant, which has all desirable traits. This starting plant is induced to produce doubled haploid from haploid cells, and later on creating homozygous/doubled haploid plants from those cells. While in natural offspring genetic recombination occurs and traits can be unlinked from each other, in doubled haploid cells and in the resulting DH plants recombination is no longer an issue. There, a recombination between two corresponding chromosomes does not lead to un-linkage of alleles or traits, since it just leads to recombination with its identical copy. Thus, traits on one chromosome stay linked. Selecting those offspring having the desired set of chromosomes and crossing them will result in a final F1 hybrid plant, having exactly the same set of chromosomes, genes and traits as the starting hybrid plant. The homozygous parental lines can reconstitute the original heterozygous plant by crossing, if desired even in a large quantity. An individual heterozygous plant can be converted into a heterozygous variety (F1 hybrid) without the necessity of vegetative propagation but as the result of the cross of two homozygous/doubled haploid lines derived from the originally selected plant. patent

Genetic Modification

Genetic modification of plants is achieved by adding a specific gene or genes to a plant, or by knocking down a gene with RNAi, to produce a desirable phenotype. The plants resulting from adding a gene are often referred to as transgenic plants. If for genetic modification genes of the species or of a crossable plant are used under control of their native promoter, then they are called cisgenic plants. Sometimes genetic modification can produce a plant with the desired trait or traits faster than classical breeding because the majority of the plant's genome is not altered.

To genetically modify a plant, a genetic construct must be designed so that the gene to be added or removed will be expressed by the plant. To do this, a promoter to drive transcription and a termination sequence to stop transcription of the new gene, and the gene or genes of interest must be introduced to the plant. A marker for the selection of transformed plants is also included. In the laboratory, antibiotic resistance is a commonly used marker: Plants that have been successfully transformed will grow on media containing antibiotics; plants that have not been transformed will die. In some instances markers for selection are removed by backcrossing with the parent plant prior to commercial release.

The construct can be inserted in the plant genome by genetic recombination using the bacteria *Agrobacterium tumefaciens* or *A. rhizogenes*, or by direct methods like the gene gun or microinjection. Using plant viruses to insert genetic constructs into plants is also a possibility, but the technique is limited by the host range of the virus. For example, Cauliflower mosaic virus (CaMV) only infects cauliflower and related species. Another limitation of viral vectors is that the virus is not usually passed on the progeny, so every plant has to be inoculated.

The majority of commercially released transgenic plants are currently limited to plants that have introduced resistance to insect pests and herbicides. Insect resistance is achieved through incorporation of a gene from *Bacillus thuringiensis* (Bt) that encodes a protein that is toxic to some insects. For example, the cotton bollworm, a common cotton pest, feeds on Bt cotton it will ingest the toxin and die. Herbicides usually work by binding to certain plant enzymes and inhibiting their action. The enzymes that the herbicide inhibits are known as the herbicides *target site*. Herbicide resistance can be engineered into crops by expressing a version of *target site* protein that is not inhibited by the herbicide. This is the method used to produce glyphosate resistant crop plants.

Genetic modification of plants that can produce pharmaceuticals (and industrial chemicals), sometimes called *pharming*, is a rather radical new area of plant breeding.

Issues and Concerns

Modern plant breeding, whether classical or through genetic engineering, comes with issues of concern, particularly with regard to food crops. The question of whether breeding can have a negative effect on nutritional value is central in this respect. Although relatively little direct research in this area has been done, there are scientific indications that, by favoring certain aspects of a plant's development, other aspects may be retarded. A study published in the *Journal of the American College of Nutrition* in 2004, entitled *Changes in USDA Food Composition Data for 43 Garden Crops, 1950 to 1999*, compared nutritional analysis of vegetables done in 1950 and in 1999, and found substantial decreases in six of 13 nutrients measured, including 6% of protein and 38% of riboflavin. Reductions in calcium, phosphorus, iron and ascorbic acid were also found. The study, conducted at the Biochemical Institute, University of Texas at Austin, concluded in summary: *"We suggest that any real declines are generally most easily explained by changes in cultivated varieties between 1950 and 1999, in which there may be trade-offs between yield and nutrient content."*

The debate surrounding genetically modified food during the 1990s peaked in 1999 in terms of media coverage and risk perception, and continues today - for example, "*Germany has thrown its weight behind a growing European mutiny over genetically modified crops by banning the planting of a widely grown pest-resistant corn variety.*" The debate encompasses the ecological impact of genetically modified plants, the safety of genetically modified food and concepts used for safety evaluation like substantial equivalence. Such concerns are not new to plant breeding. Most countries have regulatory processes in place to help ensure that new crop varieties entering the marketplace are both safe and meet farmers' needs. Examples include variety registration, seed schemes, regulatory authorizations for GM plants, etc.

Plant breeders' rights is also a major and controversial issue. Today, production of new varieties is dominated by commercial plant breeders, who seek to protect their work and collect royalties through national and international agreements based in intellectual property rights. The range of related issues is complex. In the simplest terms, critics of the increasingly restrictive regulations argue that, through a combination of technical and economic pressures, commercial breeders are reducing biodiversity and significantly constraining individuals (such as farmers) from developing and trading seed on a regional level. Efforts to strengthen breeders' rights, for example, by lengthening periods of variety protection, are ongoing.

When new plant breeds or cultivars are bred, they must be maintained and propagated. Some plants are propagated by asexual means while others are propagated by seeds. Seed propagated cultivars require specific control over seed source and production procedures to maintain the integrity of the plant breeds results. Isolation is necessary to prevent cross contamination with related plants or the mixing of seeds after harvesting. Isolation is normally accomplished by planting distance but in certain crops, plants are enclosed in greenhouses or cages (most commonly used when producing F1 hybrids.)

Role of Plant Breeding in Organic Agriculture

Critics of organic agriculture claim it is too low-yielding to be a viable alternative to conventional agriculture. However, part of that poor performance may be the result of growing poorly adapted varieties. It is estimated that over 95% of organic agriculture is based on conventionally adapted varieties, even though the production environments found in organic vs. conventional farming systems are vastly different due to their distinctive management practices. Most notably, organic farmers have fewer inputs available than conventional growers to control their production environments. Breeding varieties specifically adapted to the unique conditions of organic agriculture is critical for this sector to realize its full potential. This requires selection for traits such as:

- Water use efficiency
- Nutrient use efficiency (particularly nitrogen and phosphorus)
- Weed competitiveness
- Tolerance of mechanical weed control
- Pest/disease resistance
- Early maturity (as a mechanism for avoidance of particular stresses)
- Abiotic stress tolerance (i.e. drought, salinity, etc...)

Currently, few breeding programs are directed at organic agriculture and until recently those that did address this sector have generally relied on indirect selection (i.e. selection in conventional environments for traits considered important for organic agriculture). However, because the difference between organic and conventional environments is large, a given genotype may perform very differently in each environment due to an interaction between genes and the environment (see gene-environment interaction). If this interaction is severe enough, an important trait required for the organic environment may not be revealed in the conventional environment, which can result in the selection of poorly adapted individuals. To ensure the most adapted varieties are identified, advocates of organic breeding now promote the use of direct selection (i.e. selection in the target environment) for many agronomic traits.

There are many classical and modern breeding techniques that can be utilized for crop improvement in organic agriculture despite the ban on genetically modified organisms. For instance, controlled crosses between individuals allow desirable genetic variation to be recombined and transferred to seed progeny via natural processes. Marker assisted selection can also be employed as a diagnostics tool to facilitate selection of progeny who possess the desired trait(s), greatly speeding up the breeding process. This technique has proven particularly useful for the introgression of resistance genes into new backgrounds, as well as the efficient selection of many resistance genes pyramided into a single individual. Unfortunately, molecular markers are not currently available for many important traits, especially complex ones controlled by many genes.

Addressing Global Food Security Through Plant Breeding

For future agriculture to thrive there are necessary changes which must be made in accordance to arising global issues. These issues are arable land, harsh cropping conditions and food security which involves, being able to provide the world population with food containing sufficient nutrients. These crops need to be able to mature in several environments allowing for worldwide access, this is involves issues such as drought tolerance. These global issues are achievable through the process of plant breeding, as it offers the ability to select specific genes allowing the crop to perform at a level which yields the desired results.

Minimal Land Degradation

Land degradation is a major issue, as it can negatively impact the capability of the land to be productive. Poor agricultural management has a huge impact on the degradation of soil worldwide and it is Africa and Asia that are most affected. Through education and development of modified plants, these statistics can be reduced and agricultural land can become more productive. Plant breeding allows for an increase in yield with out the extra strain on the land. The genetically modified, Bt white maize, was introduced to South Africa and was surveyed in 33 large commercial farms and 368 small landholders properties and in both cases a higher yield was recorded.

Increased Yield Without Expansion

With an increasing population, the production of food needs to increase with it. It is estimated that a 70% increase in food production is needed by 2050 in order to meet the Declaration of the World Summit on Food Security. But with the natural degradation of agricultural land, simply planting more crops is no longer a viable option. Therefore, new varieties of plants need to be developed through plant breeding that generates an increase of yield without relying on an increase in land area. An example of this can be seen in Asia, where food production per capita has increased twofold. This has been achieved through not only the use of fertilisers, but through the use of better crops that have been specifically designed for the area.

Breeding for Increased Nutritional Value

Plant breeding can contribute to global food security as it is a cost-effective tool for increasing nutritional value of forage and crops. Improvements in nutritional value for forage crops from the use of analytical chemistry and rumen fermentation technology have been recorded since 1960; this science and technology gave breeders the ability to screen thousands of samples within a small amount of time, meaning breeders could identify a high performing hybrid quicker. The main area genetic increases were made was in vitro dry matter digestibility (IVDMD) resulting in 0.7-2.5% increase, at just 1% increase in IVDMD a single Bos Taurus also known as beef cattle reported 3.2% increase in daily gains. This improvement indicates plant breeding is an essential tool in gearing future agriculture to perform at a more advanced level.

Breeding for Tolerance

Plant breeding of hybrid crops has become extremely popular worldwide in an effort to combat the harsh environment. With long periods of drought and lack of water or nitrogen stress tolerance has become a significant part of agriculture. Plant breeders have focused on identifying crops which will ensure crops perform under these conditions; a way to achieve this is finding strains of the crop that is resistance to drought conditions with low nitrogen. It is evident from this that plant breeding is vital for future agriculture to survive as it enables farmers to produce stress resistant crops hence improving food security.

Participatory Plant Breeding

The development of agricultural science, with phenomenon like the Green Revolution arising, have left millions of farmers in developing countries, most of whom operate small farms under unstable and difficult growing conditions, in a precarious situation. The adoption of new plant varieties by this group has been hampered by the constraints of poverty and the international policies promoting an industrialized model of agriculture. Their response has been the creation of a novel and promising set of research methods collectively known as participatory plant breeding. Participatory means that farmers are more involved in the breeding process and breeding goals are defined by farmers instead of international seed companies with their large-scale breeding programs. Farmers' groups and NGOs, for example, may wish to affirm local people's rights over genetic resources, produce seeds themselves, build farmers' technical expertise, or develop new products for niche markets, like organically grown food.

References

- Suzie Key; Julian K-C Ma & Pascal MW Drake (1 June 2008). "Genetically modified plants and human health". Journal of the Royal Society of Medecine. pp. 290–298. Retrieved 11 March 2015.
- Oldeman, l (1994). "The global extent of soil degradation" (PDF). Soil resilience and sustainable land use. 32 (5967): 818–822. Retrieved 7-11-2013.
- Bänziger (2000). "Breeding for drought and nitrogen stress tolerance in maize: from theory to practice". From Theory to Practice: 7–9. Retrieved 7-11-2013.

2

Selection Methods in Plant Breeding

The chapter serves as a source to understand the methods in plant breeding. Reproduction, pollination, plant propagation and apomixis are examined in this chapter, for a better comprehension on plant breeding. The aspects elucidated in the following content are of vital importance, and provide a better understanding on plant breeding.

Selection Methods in Plant Breeding Based on Mode of Reproduction

Plant breeders use different methods depending on the mode of reproduction of crops, which include:

- Self-fertilization, where pollen from a plant will fertilise reproductive cells or ovules of the same plant
- Cross-pollination, where pollen from one plant can only fertilize a different plant
- Asexual propagation (e.g. runners from strawberry plants) where the new plant is genetically identical to its parent
- Apomixis (self-cloning), where seeds are produced asexually and the new plant is genetically identical to its parent

The mode of reproduction of a crop determines its genetic composition, which, in turn, is the deciding factor to develop suitable breeding and selection methods. Knowledge of mode of reproduction is also essential for its artificial manipulation to breed improved types. Only those breeding and selection methods are suitable for a crop which does not interfere with its natural state or ensure the maintenance of such a state. It is due to such reasons that imposition of self-fertilization on cross-pollinating crops leads to drastic reduction in their performance.

For teaching purpose, plant breeding is presented as four categories: Line breeding (autogamous crops), population breeding (allogamous crops), hybrid breeding (mostly allogamous crops, some autogamous crops), clone breeding (vegetatively propagated crops).

Self Fertilizing Crops (Autogamous Crops)

Certain restrictions caused the mechanisms for self-fertilization (partial and full self-fertilization) to develop in a number of plant species. Some of the reasons why a self-fertilizing method of reproduction is so effective are the efficacy of reproduction, as well as decreasing genetic variation and thus the fixation of highly adapted genotypes. Almost no inbreeding depression occurs in self-fertilizing plants because the mode of reproduction allows natural selection to take place in wild populations of such plants.

Critical steps in the improvement of self-fertilizing crops are the choice of parents and the identification of the best plants in segregating generations. The breeder should also have definite goals with the choice of parents. Self-fertilizing are easier to maintain, but this could lead to misuse of seed.

Some of the agronomy important, self-fertilizing crops include wheat, rice, barley, dry beans, soy beans, peanuts, tomatoes, etc.

Mass Selection

This method of selection depends mainly on selection of plants according to their phenotype and performance. The seed from selected plants are bulked for the next generation. This method is used to improve the overall population by positive or negative mass selection. Mass selection is only applied to a limited degree in self-fertilizing plants and is an effective method for the improvement of land races. This method of selection will only be effective for highly heritable traits. One shortage of mass selection are the large influence that the environment has on the development, phenotype and performance of single plants. This can also be an advantage in that varieties can be selected for local performance.

Stratified mass selection for ear size over 22 cycles has drastically altered plant phenotype in the maize population Zacatecas 58. Plants in the C22 cycle were 50 cm taller, had twice the leaf area index, reached anthesis 7 days later and had a 30% higher harvest index than Co (Table 1) Differences in growth were detected early in ontogeny. The root growth of C22 exceeded that of Co and the ratio of shoot dry mass to root dry mass was reduced by nearly 12%, from 8.0±0.2 to 7.1±0.1 (Table 2). Analysis of yield components revealed that C22 was superior to Co in grain weight, number of rows per ear, number of grains per row, and total yield per unit area (Table 3). Because the two genotypes were phenologically different, planting density optima are probably different for each population.

Selection of Cross-pollinated Crops

Plant species where normal mode of seed set is through a high degree of cross-pollination have characteristic reproductive features and population structure. Existence of self-sterility, self-incompatibility, imperfect flowers, and mechanical obstructions make the plant dependent upon foreign pollen for normal seed set. Each plant receives a blend of pollen from a large number of individuals each having different genotypes. Such populations are characterized by a high degree of heterozygosity with tremendous free and potential genetic variation, which is maintained in a steady state by free gene flow among individuals within the populations.

In the development of hybrid varieties, the aim is to identify the most productive heterozygote from the population, which then is produced with the exclusion of other members of the population.

Mass Selection

It is the simplest, easiest and oldest method of selection where individual plants are selected based on their phenotypic performance, and bulk seed selection proved to be quite effective in maize improvement at the initial stages but its efficacy especially for improvement of yield, soon came under severe criticism that culminated in the refinement of the method of mass selection. The selection after pollination does not provide any control over the pollen parent as result of which effective selection is limited only to female parents. The heritability estimates are reduced by half, since only parents are used to harvest seed whereas the pollen source is not known after the cross pollination has taken place.

Recurrent Selection

This type of selection is a refined version of the mass selection procedure and differs as follows:

- Visually selected individuals out of the base population undergo progeny testing
- Individuals selected on basis of the progeny test data are crossed with each other in every possible way to produce seed to form the new base population.

Half-sib Selection with Progeny Testing

Selections are made based on progeny test performance instead of phenotypic appearance of the parental plants. Seed from selected half-sibs, which have been pollinated by random pollen from the population (meaning that only the female parent is known and selected, hence the term "half-sib") is grown in unreplicated progeny rows for the purpose of selection. A part of the seed is planted to determine the yielding ability, or breeding value, for any character of each plant. The seed from the most productive rows or remnant seed from the outstanding half-sibs is bulked to complete one cycle of selection.

Full-sib Selection with Progeny Testing

A number of full-sib families, each produced by making crosses between the two plants from the base population are evaluated in replicated trials. A part of each full-sib family is saved for recombination. Based on evaluation the remnant seed of selected full-sib families is used to recombine the best families.

Breeding of Asexually Propagated Crops

Asexual reproduction covers all those modes of multiplication of plants where normal gamete formation and fertilization does not take place making these distinctly different from normal seed production crops. In the absence of sexual reproduction, the genetic composition of plant material being multiplied remains essentially the same as its source plant.

Clones of mother plants can be made with the exact genetic composition of the mother plant. Superior plants are selected and propagated vegetatively; the vegetative propagated offspring are used to develop stable varieties without any deterioration due to segregation of gene combinations. This unique characteristic of asexual reproduction helped to develop a number of cultivars of fruits and vegetables including grapes, apples, pears and peaches.

Improving Asexual Plant Material Through Selection

The selection in these crops is restricted to the material introduced from other sources, such as field plantations. The improvement of asexually propagated plants through induced mutations has distinct advantages and limitations. Any vegetative propagule can be treated with mutagens and even a single desirable mutant or a part of a mutated propagule (chimera) can be multiplied as an improved type of the original variety.

Selection of Asexual Plants

Selection, in the case of asexual plants, can be defined as the selection of the best performing plant and the vegetative propagation thereof. Because plants are not totally genetically stable, it can be expected that deviations would occur through the years. Selection is thus an ongoing process where deviants are selected or removed from the selection program. The main purpose of selection is to better the quality and yield of forthcoming plantations. Different approaches can be followed in the selection process of asexual plants, such as mass selection and clone selection from clone blocks.

In mass selection there are some factors that must be considered when selecting plants in a mother block, e.g. vineyard. Time of selection is a big factor, because you have to select when most of the characteristics of the plant are clearly showing. With asexual perennials the best time is just before harvest. For the best results the selected plant must be evaluated during the next season, when growth-abnormalities, leave disfigurations and virus symptoms are best visualized. Mass selection is done annually on the same plant for a minimum of three years. A plant that does not conform to the requirements in any given year of the selection cycle is discarded from the program.

New Clone Development

The development and registration of new clones take place by means of local clone selection in old plantations, as well as the importation of high quality clones from abroad, for local evaluation.

A clone is the vegetative offspring of one specific mother plant; it does not show any genetic, morphologic or physiologic deviations from the mother plant. Evaluation takes place with the different selected clones after selection.

Reproduction

Production of new individuals along a leaf margin of the miracle leaf plant (*Kalanchoe pinnata*). The small plant in front is about 1 cm (0.4 in) tall. The concept of "individual" is obviously stretched by this asexual reproductive process.

Reproduction (or procreation, breeding) is the biological process by which new individual organisms – "offspring" – are produced from their "parents". Reproduction is a fundamental feature of all known life; each individual organism exists as the result of reproduction. There are two forms of reproduction: asexual and sexual.

In asexual reproduction, an organism can reproduce without the involvement of another organism. Asexual reproduction is not limited to single-celled organisms. The cloning of an organism is a form of asexual reproduction. By asexual reproduction, an organism creates a genetically similar or identical copy of itself. The evolution of sexual reproduction is a major puzzle for biologists. The two-fold cost of sex is that only 50%

of organisms reproduce and organisms only pass on 50% of their genes.

Sexual reproduction typically requires the sexual interaction of two specialized organisms, called gametes, which contain half the number of chromosomes of normal cells and are created by meiosis, with typically a male fertilizing a female of the same species to create a fertilized zygote. This produces offspring organisms whose genetic characteristics are derived from those of the two parental organisms.

Asexual

Asexual reproduction is a process by which organisms create genetically similar or identical copies of themselves without the contribution of genetic material from another organism. Bacteria divide asexually via binary fission; viruses take control of host cells to produce more viruses; Hydras (invertebrates of the order *Hydroidea*) and yeasts are able to reproduce by budding. These organisms often do not possess different sexes, and they are capable of "splitting" themselves into two or more copies of themselves. Most plants have the ability to reproduce asexually and the ant species Mycocepurus smithii is thought to reproduce entirely by asexual means.

Some species that are capable of reproducing asexually, like hydra, yeast Mating of yeasts and jellyfish, may also reproduce sexually. For instance, most plants are capable of vegetative reproduction without seeds or spores but can also repro-duce sexually. Likewise, bacteria may exchange genetic information by conjugation.

Other ways of asexual reproduction include parthenogenesis, fragmentation and spore formation that involves only mitosis. Parthenogenesis is the growth and development of embryo or seed without fertilization by a male. Parthenogenesis occurs naturally in some species, including lower plants (where it is called apomixis), invertebrates (e.g. water fleas, aphids, some bees and parasitic wasps), and vertebrates (e.g. some reptiles, fish, and, very rarely, birds and sharks). It is sometimes also used to describe reproduction modes in hermaphroditic species which can self-fertilize.

Sexual

Sexual reproduction is a biological process that creates a new organism by combining the genetic material of two organisms in a process that starts with meiosis, a specialized type of cell division. Each of two parent organisms contributes half of the offspring's genetic makeup by creating haploid gametes. Most organisms form two different types of gametes. In these *anisogamous* species, the two sexes are referred to as male (producing sperm or microspores) and female (producing ova or megaspores). In *isogamous species*, the gametes are similar or identical in form (isogametes), but may have separable properties and then may be given other different names. For example, in the green alga, *Chlamydomonas reinhardtii*, there are so-called "plus" and "minus" gametes.

A few types of organisms, such as ciliates, *Paramecium aurelia,* have more than two types of "sex", called syngens.

Hoverflies mating in midair flight

Most animals (including humans) and plants reproduce sexually. Sexually reproducing organisms have different sets of genes for every trait (called alleles). Offspring inherit one allele for each trait from each parent, thereby ensuring that offspring have a combination of the parents' genes. Diploid having two copies of every gene within an organism, it is believed that "the masking of deleterious alleles favors the evolution of a dominant diploid phase in organisms that alternate between haploid and diploid phases" where recombination occurs freely.

Bryophyte reproduces sexually but its commonly seen life forms are all haploid, which produce gametes. The zygotes of the gametes develop into sporangium, which produces haploid spores. The diploid stage is relatively short compared with that of haploid stage, i.e. *haploid dominance*. The advantage of diploid, e.g. heterosis, only takes place in diploid life stage. Bryophyte still maintains the sexual reproduction during its evolution despite the fact that the haploid stage does not benefit from heterosis at all. This may be an example that the sexual reproduction has a bigger advantage by itself, since it allows gene shuffling (hybrid or recombination between multiple loci) among different members of the species, that permits natural selection of the fit over these new hybrids or recombinants that are haploid forms.

Allogamy

Allogamy is the fertilization of an ovum from one individual with the spermatozoa of another.

Autogamy

Self-fertilization, also known as autogamy, occurs in hermaphroditic organisms where the two gametes fused in fertilization come from the same individual, e.g., some foraminiferans, some ciliates. The term "autogamy" is sometimes substituted for autogamous pollination (not necessarily leading to successful fertilization) and describes self-pollination within the same flower, distinguished from geitonogamous pollination, transfer of pollen to a different flower on the same flowering plant, or within a single monoecious Gymnosperm plant. For example, species *Helonias bullata* suffer from low genetic diversity due to self-fertilization.

Mitosis and Meiosis

Mitosis and meiosis are types of cell division. Mitosis occurs in somatic cells, while meiosis occurs in gametes.

Mitosis The resultant number of cells in mitosis is twice the number of original cells. The number of chromosomes in the offspring cells is the same as that of the parent cell.

Meiosis The resultant number of cells is four times the number of original cells. This results in cells with half the number of chromosomes present in the parent cell. A diploid cell duplicates itself, then undergoes two divisions (tetraploid to diploid to haploid), in the process forming four haploid cells. This process occurs in two phases, meiosis I and meiosis II.

Same-sex

In recent decades, developmental biologists have been researching and developing techniques to facilitate same-sex reproduction. The obvious approaches, subject to a growing amount of activity, are female sperm and male eggs, with female sperm closer to being a reality for humans, given that Japanese scientists have already created female sperm for chickens. "However, the ratio of produced W chromosome-bearing (W-bearing) spermatozoa fell substantially below expectations. It is therefore concluded that most of the W-bearing PGC could not differentiate into spermatozoa because of restricted spermatogenesis." In 2004, by altering the function of a few genes involved with imprinting, other Japanese scientists combined two mouse eggs to produce daughter mice.

Strategies

There are a wide range of reproductive strategies employed by different species. Some animals, such as the human and northern gannet, do not reach sexual maturity for many years after birth and even then produce few offspring. Others reproduce quickly; but, under normal circumstances, most offspring do not survive to adulthood. For example, a rabbit (mature after 8 months) can produce 10–30 offspring per year, and a fruit fly (mature after 10–14 days) can produce up to 900 offspring per year. These two main strategies are known as K-selection (few offspring) and r-selection (many offspring). Which strategy is favoured by evolution depends on a variety of circumstances.

Animals with few offspring can devote more resources to the nurturing and protection of each individual offspring, thus reducing the need for many offspring. On the other hand, animals with many offspring may devote fewer resources to each individual offspring; for these types of animals it is common for many offspring to die soon after birth, but enough individuals typically survive to maintain the population. Some organisms such as honey bees and fruit flies retain sperm in a process called sperm storage thereby increasing the duration of their fertility.

Other Types

- Polycyclic animals reproduce intermittently throughout their lives.
- Semelparous organisms reproduce only once in their lifetime, such as annual plants (including all grain crops), and certain species of salmon, spider, bamboo and century plant. Often, they die shortly after reproduction. This is often associated with r-strategists.
- Iteroparous organisms produce offspring in successive (e.g. annual or seasonal) cycles, such as perennial plants. Iteroparous animals survive over multiple seasons (or periodic condition changes). This is more associated with K-strategists.

Asexual vs. Sexual Reproduction

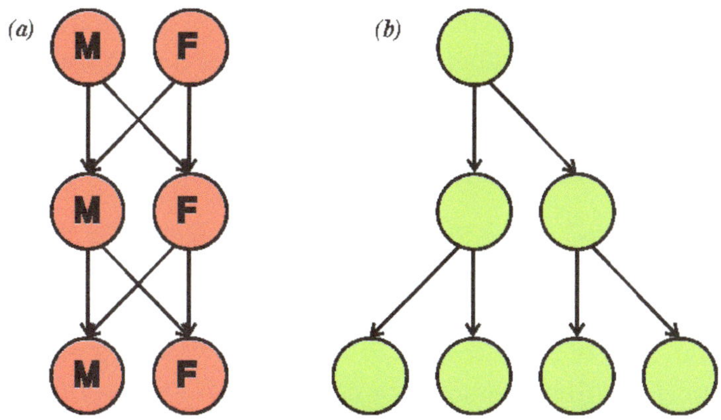

Illustration of the *twofold cost of sexual reproduction*. If each organism were to contribute to the same number of offspring (two), *(a)* the population remains the same size each generation, where the *(b)* asexual population doubles in size each generation.

Organisms that reproduce through asexual reproduction tend to grow in number exponentially. However, because they rely on mutation for variations in their DNA, all members of the species have similar vulnerabilities. Organisms that reproduce sexually yield a smaller number of offspring, but the large amount of variation in their genes makes them less susceptible to disease.

Many organisms can reproduce sexually as well as asexually. Aphids, slime molds, sea anemones, some species of starfish (by fragmentation), and many plants are examples. When environmental factors are favorable, asexual reproduction is employed to exploit suitable conditions for survival such as an abundant food supply, adequate shelter, favorable climate, disease, optimum pH or a proper mix of other lifestyle requirements. Populations of these organisms increase exponentially via asexual reproductive strategies to take full advantage of the rich supply resources.

When food sources have been depleted, the climate becomes hostile, or individual survival is jeopardized by some other adverse change in living conditions, these organisms switch to sexual forms of reproduction. Sexual reproduction ensures a mixing of the gene pool of the species. The variations found in offspring of sexual reproduction allow some individuals to be better suited for survival and provide a mechanism for selective adaptation to occur. The meiosis stage of the sexual cycle also allows especially effective repair of DNA damages Meiosis and Bernstein et al. In addition, sexual reproduc-tion usually results in the formation of a life stage that is able to endure the conditions that threaten the offspring of an asexual parent. Thus, seeds, spores, eggs, pupae, cysts or other "over-wintering" stages of sexual reproduction ensure the survival during un-favorable times and the organism can "wait out" adverse situations until a swing back to suitability occurs.

Life Without

The existence of life without reproduction is the subject of some speculation. The biological study of how the origin of life produced reproducing organisms from non-reproducing elements is called abiogenesis. Whether or not there were several independent abiogenetic events, biologists believe that the last universal ancestor to all present life on Earth lived about 3.5 billion years ago.

Scientists have speculated about the possibility of creating life non-reproductively in the laboratory. Several scientists have succeeded in producing simple viruses from entirely non-living materials. However, viruses are often regarded as not alive. Being nothing more than a bit of RNA or DNA in a protein capsule, they have no metabolism and can only replicate with the assistance of a hijacked cell's metabolic machinery.

The production of a truly living organism (e.g. a simple bacterium) with no ancestors would be a much more complex task, but may well be possible to some degree according to current biological knowledge. A synthetic genome has been transferred into an existing bacterium where it replaced the native DNA, resulting in the artificial production of a new *M. mycoides* organism.

There is some debate within the scientific community over whether this cell can be considered completely synthetic on the grounds that the chemically synthesized genome was an almost 1:1 copy of a naturally occurring genome and, the recipient cell was a naturally occurring bacterium. The Craig Venter Institute maintains the term "synthetic bacterial cell" but they also clarify "...we do not consider this to be "creating life from scratch" but rather we are creating new life out of already existing life using synthetic DNA". Venter plans to patent his experimental cells, stating that "they are pretty clearly human inventions". Its creators suggests that building 'synthetic life' would allow researchers to learn about life by building it, rather than by tearing it apart. They also propose to stretch the boundaries between life and machines until the two overlap to yield "truly programmable organisms". Researchers involved stated that the creation of "true synthetic biochemical life" is relatively close in reach with current technology and cheap compared to the effort needed to place man on the Moon.

Lottery Principle

Sexual reproduction has many drawbacks, since it requires far more energy than asexual reproduction and diverts the organisms from other pursuits, and there is some argument about why so many species use it. George C. Williams used lottery tickets as an analogy in one explanation for the widespread use of sexual reproduction. He argued that asexual reproduction, which produces little or no genetic variety in offspring, was like buying many tickets that all have the same number, limiting the chance of "winning" - that is, producing surviving offspring. Sexual reproduction, he argued, was like purchasing fewer tickets but with a greater variety of numbers and therefore a greater chance of success. The point of this analogy is that since asexual reproduction does not produce genetic variations, there is little ability to quickly adapt to a changing environment. The lottery principle is less accepted these days because of evidence that asexual reproduction is more prevalent in unstable environments, the opposite of what it predicts.

Pollination

Carpenter bee with pollen collected from Night-blooming cereus

Pollination is the process by which pollen is transferred to the female reproductive organs of a plant, thereby enabling fertilization to take place. Like all living organisms, seed plants have a single major goal: to pass their genetic information on to the next generation. The reproductive unit is the seed, and pollination is an essential step in the production of seeds in all spermatophytes (seed plants).

For the process of pollination to be successful, a pollen grain produced by the anther, the male part of a flower, must be transferred to a stigma, the female part of the flower, of a plant of the same species. The process is rather different in angiosperms (flowering plants) from what it is in gymnosperms (other seed plants). In angiosperms, after the pollen grain has landed on the stigma, it creates a pollen tube which grows down the style until it reaches the ovary. Sperm cells from the pollen grain then move along the pollen tube, enter the egg cell through the micropyle and fertilise it, resulting in the production of a seed.

Tip of a tulip stamen covered with pollen grains.

A successful angiosperm pollen grain (gametophyte) containing the male gametes is transported to the stigma, where it germinates and its pollen tube grows down the style to the ovary. Its two gametes travel down the tube to where the gametophyte(s) containing the female gametes are held within the carpel. One nucleus fuses with the polar bodies to produce the endosperm tissues, and the other with the ovule to produce the embryo Hence the term: “double fertilization”.

In gymnosperms, the ovule is not contained in a carpel, but exposed on the surface of a dedicated support organ, such as the scale of a cone, so that the penetration of carpel tissue is unnecessary. Details of the process vary according to the division of gymnosperms in question. Two main modes of fertilization are found in gymnosperms. Cycads and *Ginkgo* have motile sperm that swim directly to the egg inside the ovule, whereas conifers and gnetophytes have sperm that are unable to swim but are conveyed to the egg along a pollen tube.

The study of pollination brings together many disciplines, such as botany, horticulture, entomology, and ecology. The pollination process as an interaction between flower and pollen vector was first addressed in the 18th century by Christian Konrad Sprengel. It is important in horticulture and agriculture, because fruiting is dependent on fertilization: the result of pollination. The study of pollination by insects is known as *anthecology*.

Pollination Process

Pollen germination has three stages; hydration, activation and pollen tube emergence. The pollen grain is severely dehydrated so that its mass is reduced enabling it to be more easily transported from flower to flower. Germination only takes place after re-hydration, ensuring that premature germination does not take place in the anther. Hydration allows the plasma membrane of the pollen grain to reform into its normal bilayer organization providing an effective osmotic membrane. Activation involves the development of actin filaments throughout the cytoplasm of the cell, which eventually become concentrated at the point from which the pollen tube will emerge. Hydration and activation continue as the pollen tube begins to grow.

In conifers, the reproductive structures are borne on cones. The cones are either pollen cones (male) or ovulate cones (female), but some species are monoecious and others dioecious. A pollen cone contains hundreds of microsporangia carried on (or borne on) reproductive structures called sporophylls. Spore mother cells in the microsporangia divide by meiosis to form haploid microspores that develop further by two mitotic divisions into immature male gametophytes (pollen grains). The four resulting cells consist of a large tube cell that forms the pollen tube, a generative cell that will produce two sperm by mitosis, and two prothallial cells that degenerate. These cells comprise a very reduced microgametophyte, that is contained within the resistant wall of the pollen grain.

The pollen grains are dispersed by the wind to the female, ovulate cone that is made up of many overlapping scales (sporophylls, and thus megasporophylls), each protecting two ovules, each of which consists of a megasporangium (the nucellus) wrapped in two layers of tissue, the integument and the cupule, that were derived from highly modified branches of ancestral gymnosperms. When a pollen grain lands close enough to the tip of an ovule, it is drawn in through the micropyle (a pore in the integuments covering the tip of the ovule) often by means of a drop of liquid known as a pollination drop. The pollen enters a pollen chamber close to the nucellus, and there it may wait for a year before it germinates and forms a pollen tube that grows through the wall of the megasporangium (=nucellus) where fertilisation takes place. During this time, the megaspore mother cell divides by meiosis to form four haploid cells, three of which degenerate. The surviving one develops as a megaspore and divides repeatedly to form an immature female gametophyte (egg sac). Two or three archegonia containing an egg then develop inside the gametophyte. Meanwhile, in the spring of the second year two sperm cells are produced by mitosis of the body cell of the male gametophyte. The pollen tube elongates and pierces and grows through the megasporangium wall and delivers the sperm cells to the female gametophyte inside. Fertilisation takes place when the nucleus of one of the sperm cells enters the egg cell in the megagametophyte's archegonium.

In flowering plants, the anthers of the flower produce microspores by meiosis. These undergo mitosis to form male gametophytes, each of which contains two haploid cells. Meanwhile, the ovules produce megaspores by meiosis, further division of these form the female gametophytes, which are very strongly reduced, each consisting only of a few cells, one of which is the egg. When a pollen grain adheres to the stigma of a carpel it germinates, developing a pollen tube that grows through the tissues of the style, entering the ovule through the micropyle. When the tube reaches the egg sac, two sperm cells pass through it into the female gametophyte and fertilisation takes place.

Types

Abiotic

Abiotic pollination refers to situations where pollination is mediated without the involvement of other organisms. The most common form of abiotic pollination, anemophily, is pollination by wind. Wind pollination is very imprecise, with a minute proportion of pollen grains landing by chance on a suitable receptive stigma, the rest being wasted in the environment. This form of pollination is used by grasses, most conifers, and many deciduous trees. Hydrophily is pollination by water, and occurs in aquatic plants which release their pollen directly into the surrounding water. About 80% of all plant pollination is biotic. In gymnosperms, biotic pollination is generally incidental when it occurs, though some gymnosperms and their pollinators are mutually adapted for pollination. The best-known examples probably are members of the order Cycadales and associated species of beetles. Of the abiotically pollinated species of plant, 98% are anemophilous and 2% hydrophilous, their pollen being transported by water.

It is thought that among angiosperms, entomophily is the primitive state; this is indicated by the vestigial nectaries in the wind-pollinated *Urtica* and other plants, and the presence of fragrances in some of these plants. Of the angiosperms, grasses, sedges, rushes and catkin-bearing plants are in general wind pollinated. Other flowering plants are mostly biotic, the pollen being carried by animal vectors. However a number of plants in multiple families have secondarily adopted wind pollination in contrast to other members of their groups. Some plants are intermediate between the two pollination methods. common heather is regularly pollinated by insects, but produce clouds of pollen and some wind pollination is inevitable, and the hoary plantain is primarily wind pollinated, but is also visited by insects which pollinate it.

Melissodes desponsa covered in pollen

Biotic

Hummingbirds typically feed on red flowers

More commonly, the process of pollination requires pollinators: organisms that carry or move the pollen grains from the anther of one flower to the receptive part of the carpel or pistil (stigma) of another. This is biotic pollination. The various flower traits (and combinations thereof) that differentially attract one type of pollinator or another are known as pollination syndromes. At least 100,000 species of animal, and possibly as many as 200,000, act as pollinators of the estimated 250,000 species of flowering plants in the world. The majority of these pollinators are insects, but about 1,500 species of birds and mammals have been reported to visit flowers and may transfer pollen between them. Besides birds and bats which are the most frequent visitors, these include monkeys, lemurs, squirrels, rodents and possums.

Entomophily, pollination by insects, often occurs on plants that have developed colored petals and a strong scent to attract insects such as, bees, wasps and occasionally ants (Hymenoptera), beetles (Coleoptera), moths and butterflies (Lepidoptera), and flies (Diptera). The existence of insect pollination dates back to the dinosaur era.

In zoophily, pollination is performed by vertebrates such as birds and bats, particularly, hummingbirds, sunbirds, spiderhunters, honeyeaters, and fruit bats. Plants adapted to using bats or moths as pollinators typically have white petals and a strong scent and flower at night, whereas plants that use birds as pollinators tend to produce copious nectar and have red petals.

Insect pollinators such as honeybees (*Apis mellifera*), bumblebees (*Bombus terrestris*), and butterflies (*Thymelicus flavus*) have been observed to engage in flower constancy, which means they are more likely to transfer pollen to other conspecific plants. This can be beneficial for the pollinators, as flower constancy prevents the loss of pollen during interspecific flights and pollinators from clogging stigmas with pollen of other flower species. It also improves the probability that the pollinator will find productive flowers easily accessible and recognisable by familiar clues.

Mechanism

A European honey bee collects nectar, while pollen collects on its body.

Africanized honey bees immersed in Yellow Opuntia engelmannii Cactus Flower Pollen

Pollination can be accomplished by cross-pollination or by self-pollination:

- Cross-pollination, also called *allogamy*, occurs when pollen is delivered from the stamen of one flower to the stigma of a flower on another plant of the same species. Plants adapted for cross-pollination have several mechanisms to prevent self-pollination; the reproductive organs may be arranged in such a way that self-fertilisation is unlikely, or the stamens and carpels may mature at different times.

- Self-pollination occurs when pollen from one flower pollinates the same flower or other flowers of the same individual. It is thought to have evolved under conditions when pollinators were not reliable vectors for pollen transport, and is most often seen in short-lived annual species and plants that colonize new locations. Self-pollination may include *autogamy*, where pollen is transferred to the female part of the same flower; or *geitonogamy*, when pollen is transferred to another flower on the same plant. Plants adapted to self-fertilize often have similar stamen and carpel lengths. Plants that can pollinate themselves and produce viable offspring are called self-fertile. Plants that cannot fertilize themselves are called self-sterile, a condition which mandates cross-pollination for the production of offspring.

- *Cleistogamy*: is self-pollination that occurs before the flower opens. The pollen is released from the anther within the flower or the pollen on the anther grows a tube down the style to the ovules. It is a type of sexual breeding, in contrast to asexual systems such as apomixis. Some *cleistogamous* flowers never open, in contrast to *chasmogamous* flowers that open and are then pollinated. Cleistogamous flowers are by necessity found on self-compatible or self-fertile plants. Although certain orchids and grasses are entirely cleistogamous, other plants resort to this strategy under adverse conditions. Often there may be a mixture of both cleistogamous and chasmogamous flowers, sometimes on different parts of the plant and sometimes in mixed inflorescences. The ground bean produces cleistogamous flowers below ground, and mixed cleistogamous and chasmogamous flowers above.

Geranium incanum, like most geraniums and pelargoniums, sheds its anthers, sometimes its stamens as well, as a barrier to self-pollination. This young flower is about to open its anthers, but has not yet fully developed its pistil.

These *Geranium incanum* flowers have opened their anthers, but not yet their stigmas. Note the change of colour that signals to pollinators that it is ready for visits.

This *Geranium incanum* flower has shed its stamens, and deployed the tips of its pistil without accepting pollen from its own anthers. (It might of course still receive pollen from younger flowers on the same plant.)

An estimated 48.7% of plant species are either dioecious or self-incompatible obligate out-crossers. It is also estimated that about 42% of flowering plants exhibit a mixed mating system in nature. In the most common kind of mixed mating system, individual plants produce a single type of flower and fruits may contain self-pollinated, out-crossed or a mixture of progeny types.

Pollination also requires consideration of pollenizers. The terms "pollinator" and "pollenizer" are often confused: a *pollinator* is the agent that moves the pollen, whether it be bees, flies, bats, moths, or birds; a *pollenizer* is the plant that serves as the pollen source for other plants. Some plants are *self-compatible* (*self-fertile*) and can pollinate and fertilize themselves. Other plants have chemical or physical barriers to self-pollination.

In agriculture and horticulture pollination management, a good pollenizer is a plant that provides compatible, viable and plentiful pollen and blooms at the same time as the plant that is to be pollinated or has pollen that can be stored and used when needed to pollinate the desired flowers. Hybridization is effective pollination between flowers of different species, or between different breeding lines or populations.

Peaches are considered self-fertile because a commercial crop can be produced without cross-pollination, though cross-pollination usually gives a better crop. Apples are considered self-incompatible, because a commercial crop must be cross-pollinated. Many commercial fruit tree varieties are grafted clones, genetically identical. An orchard block of apples of one variety is genetically a single plant. Many growers now consider this a mistake. One means of correcting this mistake is to graft a limb of an appropriate pollenizer (generally a variety of crabapple) every six trees or so.

The wasp *Mischocyttarus* rotundicollis transporting pollen grains of *Schinus terebinthifolius*

Pollen Vectors

Biotic pollen vectors are animals, usually insects, but also reptiles, birds, mammals, and sundry others, that routinely transport pollen and play a role in pollination. This is usually as a result of their activities when visiting plants for feeding, breeding or shelter. The pollen adheres to the vector's body parts such as face, legs, mouthparts, hair, feathers, and moist spots; depending on the particular vector. Such transport is vital to the pollination of many plant species.

Any kind of animal that often visits or encounters flowers is likely to be a pollen vector to some extent. For example, a crab spider that stops at one flower for a time and then moves on, might carry pollen incidentally, but most pollen vectors of significant interest are those that routinely visit the flowers for some functional activity. They might feed on pollen, or plant organs, or on plant secretions such as nectar, and carry out acts of pollination on the way. Many plants bear flowers that favour certain types of pollinator over all others. This need not always be an effective strategy, because some flowers that are of such a shape that they favor pollinators that pass by their anthers and stigmata on the way to the nectar, may get robbed by ants that are small enough to bypass the normal channels, or by short-tongued bees that bite through the bases of deep corolla tubes to extract nectar at the end opposite to the anthers and stigma. Some flowers have specialized mechanisms to trap pollinators to increase effectiveness. Other flowers will attract pollinators by odor. For example, bee species such as *Euglossa cordata* are attracted to orchids this way, and it has been suggested that the bees will become intoxicated during these visits to the orchid flowers, which last up to 90 minutes. However, in general, plants that rely on pollen vectors tend to be adapted to their particular type of vector, for example day-pollinated species tend to be brightly coloured, but if they are pollinated largely by birds or specialist mammals, they tend to be larger and have larger nectar rewards than species that are strictly insect-pollinated. They also tend to spread their rewards over longer periods, having long flowering seasons; their specialist pollinators would be likely to starve if the pollination season were too short.

As for the types of pollinators, reptile pollinators are known, but they form a minority in most ecological situations. They are most frequent and most ecologically significant in island systems, where insect and sometimes also bird populations may be unstable and less species-rich. Adaptation to a lack of animal food and of predation pressure, might therefore favour reptiles becoming more herbivorous and more inclined to feed on pollen and nectar. Most species of lizards in the families that seem to be significant in pollination seem to carry pollen only incidentally, especially the larger species such as Varanidae and Iguanidae, but especially several species of the Gekkonidae are active pollinators, and so is at least one species of the Lacertidae, *Podarcis lilfordi*, which pollinates various species, but in particular is the major pollinator of *Euphorbia dendroides* on various Mediterranean islands.

Mammals are not generally thought of as pollinators, but some rodents, bats and marsupials are significant pollinators and some even specialise in such activities. In South Africa certain species of *Protea* (in particular *Protea humiflora, P. amplexicaulis, P. subulifolia, P. decurrens* and *P. cordata*) are adapted to pollination by rodents (particularly Cape Spiny Mouse, *Acomys subspinosus*) and elephant shrews (*Elephantulus* species). The flowers are borne near the ground, are yeasty smelling, not colourful, and sunbirds reject the nectar with its high xylose content. The mice apparently can digest the xylose and they eat large quantities of the pollen. In Australia pollination by flying, gliding and earthbound mammals has been demonstrated.

Examples of pollen vectors include many species of wasps, that transport pollen of many plant species, being potential or even efficient pollinators.

Evolution of Plant/pollinator Interactions

The first fossil record for abiotic pollination is from fern-like plants in the late Carboniferous period. Gymnosperms show evidence for biotic pollination as early as the Triassic period. Many fossilized pollen grains show characteristics similar to the biotically dispersed pollen today. Furthermore, the gut contents, wing structures, and mouthpart morphologies of fossilized beetles and flies suggest that they acted as early pollinators. The association between beetles and angiosperms during the early Cretaceous period led to parallel radiations of angiosperms and insects into the late Cretaceous. The evolution of nectaries in late Cretaceous flowers signals the beginning of the mutualism between hymenopterans and angiosperms.

Bees provide a good example of the mutualism that exists between hymenopterans and angiosperms. Flowers provide bees with nectar (an energy source) and pollen (a source of protein). When bees go from flower to flower collecting pollen they are also depositing pollen grains onto the flowers, thus pollinating them. While pollen and nectar, in most cases, are the most notable reward attained from flowers, bees also visit flowers for other resources such as oil, fragrance, resin and even waxes. It has been estimated that bees originated with the origin or diversification of angiosperms. In addition, cases of coevolution between bee species and flowering plants have been illustrated by specialized adaptations. For example, long legs are selected for in *Rediviva neliana*, a bee that collects oil from *Diascia capsularis*, which have long spur lengths that are selected for in order to deposit pollen on the oil-collecting bee, which in turn selects for even longer legs in *R. neliana* and again longer spur length in *D. capsularis* is selected for, thus, continually driving each other's evolution.

In Agriculture

An *Andrena* bee collects pollen among the stamens of a rose. The female carpel structure appears rough and globular to the left. The bee's stash of pollen is on its hind leg.

Bombus ignitus, a popular commercial pollinator in Japan and China

Blueberries being pollinated by bumblebees. Bumblebee hives need to be bought each year as the queens must hibernate (unlike honey bees). They are used nonetheless as they offer advantages with certain fruits as blueberries (such as the fact that they are active even at colder outdoor ambient temperature).

Well-pollinated blackberry blossom begins to develop fruit. Each incipient drupelet has its own stigma and good pollination requires the delivery of many grains of pollen to the flower so that all drupelets develop.

Pollination management is a branch of agriculture that seeks to protect and enhance present pollinators and often involves the culture and addition of pollinators in monoculture situations, such as commercial fruit orchards. The largest managed pollination event in the world is in Californian almond orchards, where nearly half (about one million hives) of the US honey bees are trucked to the almond orchards each spring. New York's apple crop requires about 30,000 hives; Maine's blueberry crop uses about 50,000 hives each year.

Bees are also brought to commercial plantings of cucumbers, squash, melons, strawberries, and many other crops. Honey bees are not the only managed pollinators: a few other species of bees are also raised as pollinators. The alfalfa leafcutter bee is an important pollinator for alfalfa seed in western United States and Canada. Bumblebees are increasingly raised and used extensively for greenhouse tomatoes and other crops.

The ecological and financial importance of natural pollination by insects to agricultural crops, improving their quality and quantity, becomes more and more appreciated and has given rise to new financial opportunities. The vicinity of a forest or wild grasslands with native pollinators near agricultural crops, such as apples, almonds or coffee can improve their yield by about 20%. The benefits of native pollinators may result in forest owners demanding payment for their contribution in the improved crop results – a simple example of the economic value of ecological services. Farmers can also raise native crops in order to promote native bee pollinator species as shown with *L. vierecki* in Delware and *L. leucozonium* in southwest Virginia.

The American Institute of Biological Sciences reports that native insect pollination saves the United States agricultural economy nearly an estimated $3.1 billion annually through natural crop production; pollination produces some $40 billion worth of products annually in the United States alone.

Pollination of food crops has become an environmental issue, due to two trends. The trend to monoculture means that greater concentrations of pollinators are needed at bloom time than ever before, yet the area is forage poor or even deadly to bees for the rest of the season. The other trend is the decline of pollinator populations, due to pesticide misuse and overuse, new diseases and parasites of bees, clearcut logging, decline of beekeeping, suburban development, removal of hedges and other habitat from farms, and public concern about bees. Widespread aerial spraying for mosquitoes due to West Nile fears is causing an acceleration of the loss of pollinators.

The US solution to the pollinator shortage, so far, has been for commercial beekeepers to become pollination contractors and to migrate. Just as the combine harvesters follow the wheat harvest from Texas to Manitoba, beekeepers follow the bloom from south to north, to provide pollination for many different crops.

In some situations, farmers or horticulturists may aim to restrict natural pollination to only permit breeding with the preferred individuals plants. This may be achieved through the use of pollination bags.

Improving Pollination in Areas with Suboptimal Bee Densities

In some instances growers' demand for beehives far exceeds the available supply. The number of managed beehives in the US has steadily declined from close to 6 million after WWII, to less than 2.5 million today. In contrast, the area dedicated to growing bee-pollinated crops has grown over 300% in the same time period. Additionally, in the past five years there has been a decline in winter managed beehives, which has reached an unprecedented rate of colony losses at near 30%. At present, there is an enormous demand for beehive rentals that cannot always be met. There is a clear need across the agricultural industry for a management tool to draw pollinators into cultivations and encourage them to preferentially visit and pollinate the flowering crop. By attracting pollinators like honeybees and increasing their foraging behavior, particularly in the center of large plots, we can increase grower returns and optimize yield from their plantings. ISCA Technologies, from Riverside California, created a semiochemical formulation called SPLAT Bloom, that modifies the behavior of honeybees, inciting them to visit flowers in every portion of the field.

Environmental Impacts

Loss of pollinators, also known as Pollinator decline (of which colony collapse disorder is perhaps the most well known) has been noticed in recent years. Observed losses would have significant economic impacts. Possible explanations for pollinator decline include habitat destruction, pesticide, parasitism/diseases, climate change and others, and many researchers believe it is the synergistic effects of these factors which are ultimately detrimental to pollinator populations.

The Structure of Plant-pollinator Networks

Wild pollinators often visit a large number of plant species and plants are visited by a large number of pollinator species. All these relations together form a network of interactions between plants and pollinators. Surprising similarities were found in the structure of networks consisting out of the interactions between plants and pollinators. This structure was found to be similar in very different ecosystems on different continents, consisting of entirely different species.

The structure of plant-pollinator networks may have large consequences for the way in which pollinator communities respond to increasingly harsh conditions. Mathematical models, examining the consequences of this network structure for the stability of pollinator communities suggest that the specific way in which plant-pollinator networks are organized minimizes competition between pollinators and may even lead to strong indirect facilitation between pollinators when conditions are harsh. This means that pollinator species together can survive under harsh conditions. But it also means that pollinator species collapse simultaneously when conditions pass a critical point. This simultaneous collapse occurs, because pollinator species depend on each other when surviving under difficult conditions.

Such a community-wide collapse, involving many pollinator species, can occur suddenly when increasingly harsh conditions pass a critical point and recovery from such a collapse might not be easy. The improvement in conditions needed for pollinators to recover, could be substantially larger than the improvement needed to return to conditions at which the pollinator community collapsed.

Plant Propagation

Gentian seedlings in a plant nursery

Plant propagation is the process of creating new plants from a variety of sources: seeds, cuttings, bulbs and other plant parts. Plant propagation can also refer to the artificial or natural dispersal of plants.

Sexual Propagation (Seed)

Seeds and spores can be used for reproduction (through e.g. sowing). Seeds are typically produced from sexual reproduction within a species, because genetic recombination has occurred. A plant grown from seeds may have different characteristics from its parents. Some species produce seeds that require special conditions to germinate, such as cold treatment. The seeds of many Australian plants and plants from southern Africa and the American west require smoke or fire to germinate. Some plant species, including many trees do not produce seeds until they reach maturity, which may take many years. Seeds can be difficult to acquire and some plants do not produce seed at all. Some plants (like certain F1/F2 hybrids and GMO plants) may produce seed, but not fertile seed. In certain cases (like with GMO's), this is done to prevent the accidental spreading of these plants (which are generally non-native crops), for example by birds and other animals.

One way to propagate an avocado seed

Asexual Propagation

Plants have a number of mechanisms for asexual or vegetative reproduction. Some of these have been taken advantage of by horticulturists and gardeners to multiply or clone plants rapidly. People also use methods that plants do not use, such as tissue culture and grafting. Plants are produced using material from a single parent and as such there is no exchange of genetic material, therefore vegetative propagation methods almost always produce plants that are identical to the parent. Vegetative reproduction uses plants parts such as roots, stems and leaves. In some plants seeds can be produced without fertilization and the seeds contain only the genetic material of the parent plant. Therefore, propagation via asexual seeds or apomixis is asexual reproduction but not vegetative propagation.

Softwood stemcuttings rooting in a controlled environment

Techniques for vegetative propagation include:

- Air or ground layering
- Division
- Grafting and bud grafting, widely used in fruit tree propagation
- Micropropagation
- Stolons or runners
- Storage organs such as bulbs, corms, tubers and rhizomes
- Striking or cuttings
- Twin-scaling
- Offsets

Heated Propagator

A heated propagator is a horticultural device to maintain a warm and damp environment for seeds and cuttings to grow in.

This can be in the form of a clear enclosed bin sitting over a hotpad, or even a portable heater pointed at the bin. The key is to keep the moisture in the clear bin, while keeping lighting over the top of it, usually.

Seed Propagation Mat

An electric seed-propagation mat is a heated rubber mat covered by a metal cage which is used in gardening. The mats are made so that planters containing seedlings can be placed on top of the metal cage without the risk of starting a fire. In extreme cold, gardeners place a loose plastic cover over the planters/mats which creates a sort of miniature greenhouse. The constant and predictable heat allows people to garden in the winter months when the weather is generally too cold for seedlings to survive naturally. When combined with a lighting system, many plants can be grown indoors using these mats.

Apomixis

In botany, apomixis was defined by Hans Winkler as replacement of the normal sexual reproduction by asexual reproduction, without fertilization. Its etymology is Greek for "away from" + "mixing". This definition notably does not mention meiosis. Thus "normal asexual reproduction" of plants, such as propagation from cuttings or leaves, has never been considered to be apomixis, but replacement of the seed by a plantlet or replacement of the flower by bulbils are types of apomixis. Apomictically produced offspring are genetically identical to the parent plant.

Vegetative apomixis in *Poa bulbosa*; bulbils form instead of flowers

Some authors included all forms of asexual reproduction within apomixis, but that generalization of the term has since died out.

In flowering plants, the term "apomixis" is commonly used in a restricted sense to mean agamospermy, i.e. clonal reproduction through seeds. Although agamospermy could theoretically occur in gymnosperms, it appears to be absent in that group.

Apogamy is a related term that has had various meanings over time. In plants with independent gametophytes (notably ferns), the term is still used interchangeably with "apomixis", and both refer to the formation of sporophytes by parthenogenesis of gametophyte cells.

Male apomixis (paternal apomixis) involves replacement of the genetic material of the egg cell by that from the pollen.

Apomixis and Evolution

Because apomictic plants are genetically identical from one generation to the next, each lineage has some of the characters of a true species, maintaining distinctions from other apomictic lineages within the same genus, while having much smaller differences than is normal between species of most genera. They are therefore often called microspecies. In some genera, it is possible to identify and name hundreds or even thousands of microspecies, which may be grouped together as species aggregates, typically listed in floras with the convention "*Genus species* agg." (such as the bramble, *Rubus fruticosus* agg.). In some plant families, genera with apomixis are quite common, for example in Asteraceae, Poaceae, and Rosaceae. Examples of apomixis can be found in the genera *Crataegus* (hawthorns), *Amelanchier* (shadbush), *Sorbus* (rowans and whitebeams), *Rubus* (brambles or blackberries), *Poa* (meadow grasses), *Nardus stricta* (Matgrass), *Hieracium* (hawkweeds) and *Taraxacum* (dandelions).

Although the evolutionary advantages of sexual reproduction are lost, apomixis can pass along traits fortuitous for evolutionary fitness. As Jens Clausen put it

The apomicts actually have discovered the effectiveness of mass production long before Mr Henry Ford applied it to the production of the automobile. ... Facultative apomixis ... does not prevent variation; rather, it multiplies certain varietal products.

Facultative apomixis means that apomixis does not always occur, i.e. sexual reproduction can also happen. It appears likely that all apomixis in plants is facultative; in other words, that "obligate apomixis" is an artifact of insufficient observation (missing uncommon sexual reproduction).

Apogamy and Apospory in Non-flowering Plants

The gametophytes of bryophytes, and less commonly ferns and lycopods can develop a group of cells that grow to look like a sporophyte of the species but with the ploidy level of the gametophyte, a phenomenon known as apogamy. The sporophytes of plants of these groups may also have the ability to form a plant that looks like a gametophyte but with the ploidy level of the sporophyte, a phenomenon known as apospory.

Apomixis in Flowering Plants (Angiosperms)

Agamospermy, asexual reproduction through seeds, occurs in flowering plants through many different mechanisms and a simple hierarchical classification of the different types is not possible. Consequently, there are almost as many different usages of terminology for apomixis in angiosperms as there are authors on the subject. For English speakers, Maheshwari 1950 is very influential. German speakers might prefer to consult Rutishauser 1967. Some older text books on the basis of misinformation (that the egg cell in a meiotically unreduced gametophyte can never be fertilized) attempted to reform the terminology to match parthenogenesis as it is used in zoology, and this continues to cause much confusion.

Agamospermy occurs mainly in two forms: In *gametophytic apomixis*, the embryo arises from an unfertilized egg cell (i.e. by parthenogenesis) in a gametophyte that was produced from a cell that did not complete meiosis. In *adventitious embryony* (sporophytic apomixis), an embryo is formed directly (not from a gametophyte) from nucellus or integument tissue.

Types of Apomixis in Flowering Plants

Caribbean Agave producing plantlets on the old flower stem.

Maheshwari used the following simple classification of types of apomixis in flowering plants:

- Nonrecurrent apomixis: In this type "the megaspore mother cell undergoes the usual meiotic divisions and a haploid embryo sac [megagametophyte] is formed. The new embryo may then arise either from the egg (haploid parthenogenesis) or from some other cell of the gametophyte (haploid apogamy)." The haploid plants have half as many chromosomes as the mother plant, and "the process is not repeated from one generation to another" (which is why it is called nonrecurrent).
- Recurrent apomixis, is now more often called gametophytic apomixis: In this type, the megagametophyte has the same number of chromosomes as the mother plant because meiosis was not completed. It generally arises either from an archesporial cell or from some other part of the nucellus.
- Adventive embryony, also called sporophytic apomixis, sporophytic budding, or nucellar embryony: Here there may be a megagametophyte in the ovule, but the embryos do not arise from the cells of the gametophyte; they arise from cells of nucellus or the integument. Adventive embryony is important in several species of *Citrus*, in *Garcinia*, *Euphorbia dulcis*, *Mangifera indica* etc.
- Vegetative apomixis: In this type "the flowers are replaced by bulbils or other vegetative propagules which frequently germinate while still on the plant". Vegetative apomixis is important in *Allium*, *Fragaria*, *Agave*, and some grasses, among others.

Types of Gametophytic Apomixis

Gametophytic apomixis in flowering plants develops in several different ways. A megagametophyte develops with an egg cell within it that develops into an embryo through parthenogenesis. The central cell of the megagametophyte may require fertilization to form the endosperm, pseudogamous gametophytic apomixis, or in autonomous gametophytic apomixis fertilization is not required.

- In diplospory (also called generative apospory), the megagametophyte arises from a cell of the archesporium.
- In apospory (also called somatic apospory), the megagametophyte arises from some other nucellus cell.

Considerable confusion has resulted because diplospory is often defined to involve the megaspore mother cell only, but a number of plant families have a multicellular archesporium and the megagametophyte could originate from another archesporium cell.

Diplospory is further subdivided according to how the megagametophyte forms:

- *Allium odorum–A. nutans* type. The chromosomes double (endomitosis) and then meiosis proceeds in an unusual way, with the chromosome copies pairing up (rather than the original maternal and paternal copies pairing up).
- *Taraxacum* type: Meiosis I fails to complete, meiosis II creates two cells, one of which degenerates; three mitotic divisions form the megagametophyte.
- *Ixeris* type: Meiosis I fails to complete; three rounds of nuclear division occur without cell-wall formation; wall formation then occurs.
- *Blumea–Elymus* types: A mitotic division is followed by degeneration of one cell; three mitotic divisions form the megagametophyte.
- *Antennaria–Hieracium* types: three mitotic divisions form the megagametophyte.
- *Eragrostis–Panicum* types: Two mitotic division give a 4-nucleate megagametophyte, with cell walls to form either three or four cells.

Incidence of Apomixis in Flowering Plants

Apomixis occurs in at least 33 families of flowering plants, and has evolved multiple times from sexual relatives. Apomictic species or individual plants often have a hybrid origin, and are usually polyploid.

In plants with both apomictic and meiotic embryology, the proportion of the different types can differ at different times of year, and photoperiod can also change the proportion. It appears unlikely that there are any truly completely apomictic plants, as low rates of sexual reproduction have been found in several species that were previously thought to be entirely apomictic.

The genetic control of apomixis can involve a single genetic change that affects all the major developmental components, formation of the megagametophyte, parthenogenesis of the egg cell, and endosperm development. However, the timing of the various developmental processes is critical to successful development of an apomictic seed, and the timing can be affected by multiple genetic factors.

Some Terms Related to Apomixis

- Apomeiosis: "Without meiosis"; usually meaning the production of a meiotically unreduced gametophyte.
- Parthenogenesis: Development of an embryo directly from an egg cell without fertilization is called parthenogenesis. It is of two types:
    - Haploid parthenogenesis: Parthenogenesis of a normal haploid egg (a meiotically reduced egg) into an embryo is termed haploid parthenogenesis. If the mother plant was diploid, then the haploid embryo that results is monoploid, and the plant that grows from the embryo is sterile. If they are not sterile, they are sometimes useful to plant breeders. This type of apomixis has been recorded in *Solanum nigrum*, *Lilium* spp., *Orchis maculata*, *Nicotiana tabacum*, etc.
    - Diploid parthenogenesis: When the megagametophyte develops without completing meiosis, so that the megagametophyte and all cells within it are meiotically unreduced (a.k.a. diploid, but diploid is an ambiguous term), this is called diploid parthenogenesis, and the plant that develops from the embryo will have the same number of chromosomes as the mother plant. Diploid parthenogenesis is a component process of gametophytic apomixis.
- Androgenesis and androclinesis are synonyms. These terms are used for two different processes that both have the effect of producing an embryo that has "male inheritance".

  The first process is a natural one. It may also be referred to as male apomixis or paternal apomixis. It involves fusion of the male and female gametes and replacement of the female nucleus by the male nucleus. This has been noted as a rare phenomenon in many plants (e.g. *Nicotiana* and *Crepis*), and occurs as the regular reproductive method in the Saharan Cypress, *Cupressus dupreziana.*

  The second process that is referred to as androgenesis or androclinesis involves (artificial) culture of haploid plants from anther tissue or microspores.
- Apogamy: Although this term was (before 1908) used for other types of apomixis, and then discarded as too confusing, it is still sometimes used when an embryo develops from a cell of the megagametophyte other than the egg cell. In flowering plants, the cells involved in apogamy would be synergids or antipodal cells.

- Addition hybrids, called B_{III} hybrids by Rutishauser: An embryo is formed after a meiotically unreduced egg cell is fertilized. The ploidy level of the embryo is therefore higher than that of the mother plant. This process occurs in some plants that are otherwise apomictic, and may play a significant role in producing tetraploid plants from triploid apomictic mother plants (if they receive pollen from diploids). Because fertilization is involved, this process does not fit the definition of apomixis.

- Pseudogamy refers to any reproductive process that requires pollination but does not involve male inheritance. It is sometimes used in a restrictive sense to refer to types of apomixis in which the endosperm is fertilized but the embryo is not. A better term for the restrictive sense is centrogamy.

- Agamospecies, the concept introduced by Göte Turesson: "an apomict population the constituents of which, for morphological, cytological or other reasons, are to be considered as having a common origin," i.e., basically synonymous with "microspecies.

References

- Raghavan, Valayamghat (1997). Molecular Embryology of Flowering Plants. Cambridge University Press. pp. 210–216. ISBN 978-0-521-55246-2.

- Campbell, Neil A.; Reece, Jane B. (2002). Biology (6th edition). Pearson Education. pp. 600–612. ISBN 978-0-201-75054-6.

- Cronk, J. K.; Fennessy, M. Siobhan (2001). Wetland plants: biology and ecology. Boca Raton, Fla.: Lewis Publishers. p. 166. ISBN 1-56670-372-7.

- Glover, Beverly J. (2007). Understanding flowers and flowering: an integrated approach. Oxford University Press. p. 127. ISBN 0-19-856596-8.

- Baskin, Carol C.; Baskin, Jerry M. (2001). Seeds: Ecology, Biogeography, and Evolution of Dormancy and Germination. Elsevier. p. 215. ISBN 978-0-12-080263-0.

- Nowicki, Marcin; et al. (26 October 2013), More than meets the eye: A multi-year expressivity analyses of tomato sterility in ps and ps-2 lines (PDF), Australian Journal of Crop Science, 7(13):2154-2161; Southern Cross Publishing, retrieved 2013-10-29

3

Processes of Plant Breeding

Processes like molecular breeding, marker-assisted selection, doubled haploidy, genetic engineering, family based QTL mapping etc. are described in the following chapter in a critical and systematic manner. The following chapter elucidates the various processes and technologies that are related to plant breeding.

Molecular Breeding

Molecular breeding is the application of molecular biology tools, often in plant breeding.

The areas of molecular breeding include:

- QTL mapping or gene discovery
- Marker assisted selection and genomic selection
- Genetic engineering
- Genetic transformation

Steps

Marker Assisted Breeding

Genotyping and creating molecular maps - *genomics*

> The commonly used markers include Simple sequence repeats (or microsatellites), single nucleotide polymorphisms (SNP). The process of identification of plant genotypes is known as genotyping.

Development of SNPs has revolutionized the molecular breeding process as it help to create dense markers. Another area that is developing is genotyping by sequencing.

Phenotyping - Phenomics

> To identify genes associated with traits, it is important to measure trait value - known as phenotype. "omics" for measurement of phenotypes is called phenomics. The phenotype can be indicative of measurement of trait itself or an indirectly related or correlated trait.

QTL Mapping or Association Mapping

Genes (Quantitative trait loci (abbreviated as QTLs) or quantitative trait genes or minor genes or major genes) involved in controlling trait of interest is identified. The process is known as mapping. Mapping of such genes can be done using molecular markers. QTL mapping can involve single large family, unrelated individuals or multiple families Family based QTL mapping). Basic idea is to identify genes or markers associated with genes that can be used in marker assisted breeding / selection.

Marker Assisted Selection or Genetic Selection

Once genes or markers are identified they can be used for genotyping and selection decision can be made.

Marker-assisted Backcrossing (MABC)

Backcross is crossing F1 with its parents to transfer a limited number of loci (e.g. transgene, disease resistance loci, etc.) from one genetic background to another. Usually the recipient of such genes is good adopted cultivars otherwise except the gene is to be transferred. So we want to keep genetic background of the recipient genotypes, which is done by 4-6 rounds of repeated backcrosses while selecting for the gene of interest. We can use markers from the whole genome to recover the genome quickly in 2-3 rounds of backcrossing might be good enough in such situation.

Marker-assisted Recurrent Selection (MARS)

MARS include identification and selection of several genomic regions (up to 20 or even more) for complex traits within a single population.

Genomic Selection

Genomic selection is novel approach to traditional marker-assisted selection where selection are made based on few markers. Rather than seeking to identify individual loci significantly associated with a trait, genomics uses all marker data as predictors of performance and consequently delivers more accurate predictions. Selection can be based on genomic selection predictions, potentially leading to more rapid and lower cost gains from breeding.Genomic prediction combines marker data with phenotypic and pedigree data (when available) in an attempt to increase the accuracy of the prediction of breeding and genotypic values.

Genetic Transformation or Genetic Engineering

Transfer of genes make it possible for horizontal transfer of genes from one organism to another. Thus plants can receive genes from humans or algae or any other organism. This provides limitless opportunity in breeding crop plants.

Molecular Breeding Areas

Quantitative Trait Locus

A quantitative trait locus (QTL) is a section of DNA (the locus) that correlates with variation in a phenotype (the quantitative trait). The QTL typically is linked to, or contains, the genes that control that phenotype. QTLs are mapped by identifying which molecular markers (such as SNPs or AFLPs) correlate with an observed trait. This is often an early step in identifying and sequencing the actual genes that cause the trait variation.

Quantitative traits are phenotypes (characteristics) that vary in degree and can be attributed to polygenic effects, i.e., the product of two or more genes, and their environment.

History

Mendelian inheritance was rediscovered at the beginning of the 20th century, and as Mendel's ideas spread geneticists began to connect Mendel's rules of inheritance of single factors to Darwinian evolution. For early geneticists, it was not immediately clear that the smooth variation in traits like body size (i.e., Incomplete Dominance) was caused by the inheritance of single genetic factors. Although Darwin himself observed that inbred features of fancy pigeons were inherited in accordance with Mendel's laws (although Darwin didn't actually know about Mendel's ideas when he made the observation), it was not obvious that these features selected by fancy pigeon breeders can similarly explain quantitative variation in nature.

An early attempt by William Ernest Castle to unify the laws of mendelian inheritance with Darwin's theory of speciation invoked the idea that species become distinct from one another as one species or the other acquires a novel mendelian factor. Castle's conclusion was based on the observation that novel traits that could be studied in the lab and that show mendelian inheritance patterns reflect a large deviation from the wild type, and Castle believed that acquisition of such features is the basis of "discontinuous variation" that characterizes speciation. Interestingly, Darwin discussed the inheritance of similar mutant features but did not invoke them as a requirement of speciation. Instead Darwin used the emergence of such features in breeding populations as evidence that mutation can occur at random within breeding populations, which is a central premise of his model of selection in nature. Later in his career, Castle would refine his model for speciation to allow for small variation to contribute to speciation over time. He also was able to demonstrate this point by selectively breeding laboratory populations of rats to obtain a hooded phenotype over several generations.

Castle's was perhaps the first attempt made in the scientific literature to direct evolution by artificial selection of a trait with continuous underlying variation, however the practice had previously been widely employed in the development of agriculture to obtain livestock or plants with favorable features from populations that show quantitative variation in traits like body size or grain yield.

Castle's work was among the first to attempt to unify the recently rediscovered laws of mendelian inheritance with Darwin's theory of evolution. Still, it would be almost thirty years until the theoretical framework for evolution of complex traits would be widely formalized. In an early summary of the theory of evolution of continuous variation, Sewall Wright, a graduate student who trained under Castle, summarized contemporary thinking about the genetic basis of quantitative natural variation: "As genetic studies continued, ever smaller differences were found to mendelize, and any character, sufficiently investigated, turned out to be affected by many factors." Wright and others formalized population genetics theory that had been worked out over the preceding 30 years explaining how such traits can be inherited and create stably breeding populations with unique characteristics. Quantitative trait genetics today leverages Wright's observations about the statistical relationship between genotype and phenotype in families and populations to understand how certain genetic features can affect variation in natural and derived populations.

Quantitative Traits

Polygenic inheritance refers to inheritance of a phenotypic characteristic (trait) that is attributable to two or more genes and can be measured quantitatively. Multifactorial inheritance refers to polygenic inheritance that also includes interactions with the environment. Unlike monogenic traits, polygenic traits do not follow patterns of Mendelian inheritance (discrete categories). Instead, their phenotypes typically vary along a continuous gradient depicted by a bell curve.

An example of a polygenic trait is human skin color variation. Several genes factor into determining a person's natural skin color, so modifying only one of those genes can change skin color slightly or in some cases, such as for SLC24A5, moderately. Many disorders with genetic components are polygenic, including autism, cancer, diabetes and numerous others. Most phenotypic characteristics are the result of the interaction of multiple genes.

Examples of disease processes generally considered to be results of *multifactorial etiology*:

Congenital Malformation

- Cleft palate
- Congenital dislocation of the hip

- Congenital heart defects
- Neural tube defects
- Pyloric stenosis
- Talipes

Adult Onset Diseases

- Diabetes Mellitus
- Cancer
- *DiIorio C, Escoffery C, Yeager KA, McCarty F, Henry TR, Koganti A, Reisinger E, Robinson E, Kobau R, Price P (2009). "WebEase: development of a Web-based epilepsy self-management intervention". Prev Chronic Dis. 6: A28. PMC 2644585. PMID 19080034.*
- Glaucoma
- Hypertension
- Ischaemic heart disease
- Manic depression
- Schizophrenia
- Psoriasis
- Thyroid diseases
- Alzheimer's Disease

Multifactorially inherited diseases are said to constitute the majority of genetic disorders affecting humans which will result in hospitalization or special care of some kind.

Multifactorial Traits in General

Trait controlled by the both environment and genetic factor. Usually, multifactorial traits outside of illness result in what we see as continuous characteristics in organisms, especially human organisms such as: height, skin color, and body mass. All of these phenotypes are complicated by a great deal of give-and-take between genes and environmental effects. The continuous distribution of traits such as height and skin color described above, reflects the action of genes that do not manifest typical patterns of dominance and recessiveness. Instead the contributions of each involved locus are thought to be additive. Writers have distinguished this kind of inheritance as *polygenic*, or *quantitative inheritance*.

Thus, due to the nature of polygenic traits, inheritance will not follow the same pattern as a simple monohybrid or dihybrid cross. Polygenic inheritance can be explained as Mendelian inheritance at many loci, resulting in a trait which is normally-distributed. If n is the number of involved loci, then the coefficients of the binomial expansion of $(a + b)^{2n}$ will give the frequency of distribution of all n allele combinations. For a sufficiently high n, this binomial distribution will begin to resemble a normal distribution. From this viewpoint, a disease state will become apparent at one of the tails of the distribution, past some threshold value. Disease states of increasing severity will be expected the further one goes past the threshold and away from the mean.

Heritable Disease and Multifactorial Inheritance

A mutation resulting in a disease state is often recessive, so both alleles must be mutant in order for the disease to be expressed phenotypically. A disease or syndrome may also be the result of the expression of mutant alleles at more than one locus. When more than one gene is involved, with or without the presence of environmental triggers, we say that the disease is the result of multifactorial inheritance.

The more genes involved in the cross, the more the distribution of the genotypes will resemble a normal, or Gaussian distribution. This shows that multifactorial inheritance is polygenic, and genetic frequencies can be predicted by way of a polyhybrid Mendelian cross. Phenotypic frequencies are a different matter, especially if they are complicated by environmental factors.

The paradigm of polygenic inheritance as being used to define multifactorial disease has encountered much disagreement. Turnpenny (2004) discusses how simple polygenic inheritance cannot explain some diseases such as the onset of Type I diabetes mellitus, and that in cases such as these, not all genes are thought to make an equal contribution.

The assumption of polygenic inheritance is that all involved loci make an equal contribution to the symptoms of the disease. This should result in a normal curve distribution of genotypes. When it does not, the idea of polygenetic inheritance cannot be supported for that illness.

A Cursory Look at Some Examples

Examples of such diseases are not new to medicine. The above examples are well-known examples of diseases having both genetic and environmental components. Other examples involve atopic diseases such as eczema or dermatitis;spina bifida (open spine), and anencephaly (open skull).

While schizophrenia is widely believed to be multifactorially genetic by biopsychiatrists, no characteristic genetic markers have been determined with any certainty.

If it is shown that the brothers and sisters of the patient have the disease, then there is a strong chance that the disease is genetic and that the patient will also be a genetic carrier. This is not quite enough as it also needs to be proven that the pattern of inheritance is non-Mendelian. This would require studying dozens, even hundreds of different family pedigrees before a conclusion of multifactorial inheritance is drawn. This often takes several years.

If multifactorial inheritance is indeed the case, then the chance of the patient contracting the disease is reduced only if cousins and more distant relatives have the disease. It must be stated that while multifactorially-inherited diseases tend to run in families, inheritance will not follow the same pattern as a simple monohybrid or dihybrid cross.

If a genetic cause is suspected and little else is known about the illness, then it remains to be seen exactly how many genes are involved in the phenotypic expression of the disease. Once that is determined, the question must be answered: if two people have the required genes, why are there differences in expression between them? Generally, what makes the two individuals different are likely to be environmental factors. Due to the involved nature of genetic investigations needed to determine such inheritance patterns, this is not usually the first avenue of investigation one would choose to determine etiology.

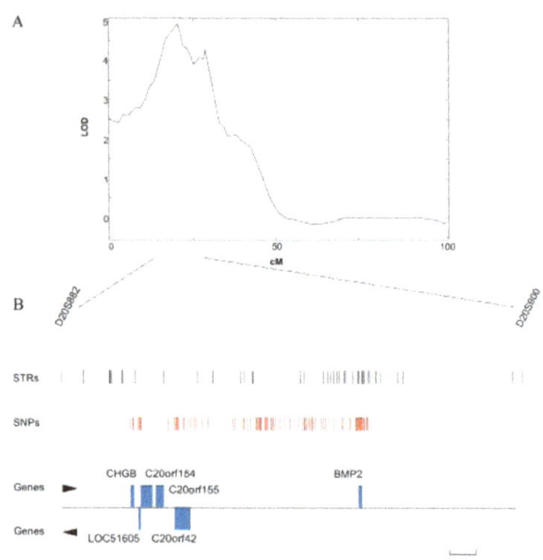

A QTL for osteoporosis on the human chromosome 20

Typically, QTLs underlie continuous traits (those traits that vary continuously, e.g. height) as opposed to discrete traits (traits that have two or several character values, e.g. red hair in humans, a recessive trait, or smooth vs. wrinkled peas used by Mendel in his experiments).

Moreover, a single phenotypic trait is usually determined by many genes. Consequently, many QTLs are associated with a single trait.

A quantitative trait locus (QTL) is a region of DNA that is associated with a particular phenotypic trait. These QTLs are often found on different chromosomes. Knowing the number of QTLs that explains variation in the phenotypic trait tells us about the genetic architecture of a trait. It may tell us that plant height is controlled by many genes of small effect, or by a few genes of large effect.

Another use of QTLs is to identify candidate genes underlying a trait. Once a region of DNA is identified as contributing to a phenotype, it can be sequenced. The DNA sequence of any genes in this region can then be compared to a database of DNA for genes whose function is already known.

In a recent development, classical QTL analyses are combined with gene expression profiling i.e. by DNA microarrays. Such expression QTLs (eQTLs) describe cis- and trans-controlling elements for the expression of often disease-associated genes. Observed epistatic effects have been found beneficial to identify the gene responsible by a cross-validation of genes within the interacting loci with metabolic pathway- and scientific literature databases.

QTL Mapping

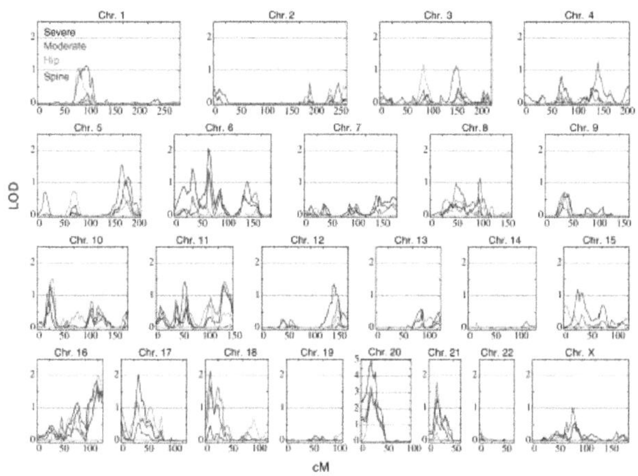

Example of a genome-wide scan for QTL of osteoporosis

For organisms whose genomes are known, one might now try to exclude genes in the identified region whose function is known with some certainty not to be connected with the trait in question. If the genome is not available, it may be an option to sequence the identified region and determine the putative functions of genes by their similarity to genes with known function, usually in other genomes. This can be done using BLAST, an online tool that allows users to enter a primary sequence and search for similar sequences within the BLAST database of genes from various organisms. It is often not the actual gene underlying the phenotypic trait, but rather a region of DNA that is closely linked with the gene.

Another interest of statistical geneticists using QTL mapping is to determine the complexity of the genetic architecture underlying a phenotypic trait. For example, they may be interested in knowing whether a phenotype is shaped by many independent loci, or by a few loci, and do those loci interact. This can provide information on how the phenotype may be evolving.

Analysis of Variance

The simplest method for QTL mapping is analysis of variance (ANOVA, sometimes called "marker regression") at the marker loci. In this method, in a backcross, one may calculate a t-statistic to compare the averages of the two marker genotype groups. For other types of crosses (such as the intercross), where there are more than two possible genotypes, one uses a more general form of ANOVA, which provides a so-called F-statistic. The ANOVA approach for QTL mapping has three important weaknesses. First, we do not receive separate estimates of QTL location and QTL effect. QTL location is indicated only by looking at which markers give the greatest differences between genotype group averages, and the apparent QTL effect at a marker will be smaller than the true QTL effect as a result of recombination between the marker and the QTL. Second, we must discard individuals whose genotypes are missing at the marker. Third, when the markers are widely spaced, the QTL may be quite far from all markers, and so the power for QTL detection will decrease.

Interval Mapping

Lander and Botstein developed interval mapping, which overcomes the three disadvantages of analysis of variance at marker loci. Interval mapping is currently the most popular approach for QTL mapping in experimental crosses. The method makes use of a genetic map of the typed markers, and, like analysis of variance, assumes the presence of a single QTL. In interval mapping, each locus is considered one at a time and the logarithm of the odds ratio (LOD score) is calculated for the model that the given locus is a true QTL. The odds ratio is related to the Pearson correlation coefficient between the phenotype and the marker genotype for each individual in the experimental cross.

The term 'interval mapping' is used for estimating the position of a QTL within two markers (often indicated as 'marker-bracket'). Interval mapping is originally based on the maximum likelihood but there are also very good approximations possible with simple regression.

The principle for QTL mapping is: 1) The Likelihood can be calculated for a given set of parameters (particularly QTL effect and QTL position) given the observed data on phenotypes and marker genotypes. 2) The estimates for the parameters are those where the likelihood are highest. 3) A significance threshold can be established by permutation testing.

CONVENTIONAL methods for the detection of quantitative trait loci (QTLs) are based on a comparison of single QTL models with a model assuming no QTL. For instance in the "interval mapping" method the likelihood for a single putative QTL is assessed at each location on the genome. However, QTLs located elsewhere on the genome can have an interfering effect. As a consequence, the power of detection may be compromised, and the estimates of locations and effects of QTLs may be biased (LANDER and BOTSTEIN 1989; KNAPP 1991). Even nonexisting so-called "ghost" QTLs may appear (HALEY and KNOTT 1992; MARTINEZ and CURNOW 1992). Therefore, it is obvious that multiple QTLs could be mapped more efficiently and more accurately by using multiple QTL models. One popular approach to handle QTL mapping where multiple QTL contribute to a trait is to iteratively scan the genome and add known QTL to the regression model as QTLs are identified. This method, termed composite interval mapping determine both the location and effects size of QTL more accurately than single-QTL approaches, especially in small mapping populations where the effect of correlation between genotypes in the mapping population may be problematic.

Composite Interval Mapping (CIM)

In this method, one performs interval mapping using a subset of marker loci as covariates. These markers serve as proxies for other QTLs to increase the resolution of interval mapping, by accounting for linked QTLs and reducing the residual variation. The key problem with CIM concerns the choice of suitable marker loci to serve as covariates; once these have been chosen, CIM turns the model selection problem into a single-dimensional scan. The choice of marker covariates has not been solved, however. Not surprisingly, the appropriate markers are those closest to the true QTLs, and so if one could find these, the QTL mapping problem would be complete anyway.

Family-pedigree Based Mapping

Family based QTL mapping, or Family-pedigree based mapping (Linkage and association mapping), involves multiple families instead of a single family. Family based QTL mapping has been the only way for mapping of genes where experimental crosses are difficult to make. However, due to some advantages, now plant geneticists are attempting to incorporate some of the methods pioneered in human genetics. Using family-pedigree based approach has been discussed (Bink et al. 2008). Family-based linkage and association has been successfully implemented (Rosyara et al. 2009)

Transformation (Genetics)

In molecular biology, transformation is the genetic alteration of a cell resulting from the direct uptake and incorporation of exogenous genetic material (exogenous DNA) from its surroundings through the cell membrane(s). Transformation occurs naturally in some species of bacteria, but in biotechnology transformation is a basic technique, and can be made to occur in many kinds of cells. For transformation to happen in nature, bacteria must be in a state of competence, which might occur as a time-limited response to environmental conditions such as starvation and cell density.

Transformation is one of three processes by which exogenous genetic material may be introduced into a bacterial cell, the other two being conjugation (transfer of genetic material between two bacterial cells in direct contact) and transduction (injection of foreign DNA by a bacteriophage virus into the host bacterium).

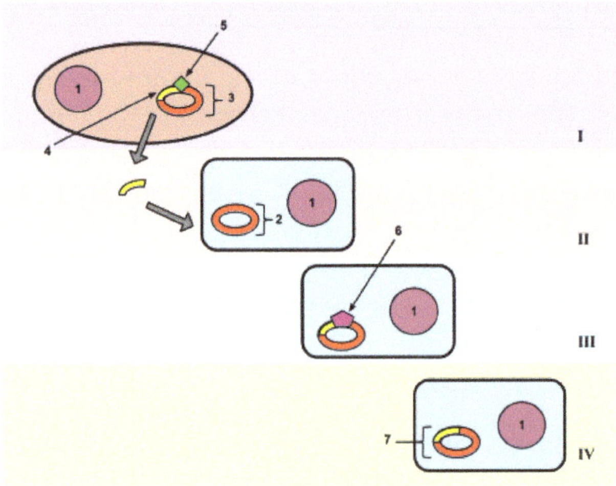

In this image, a gene from bacterial cell 1 is moved from bacterial cell 1 to bacterial cell 2. This process of bacterial cell 2 taking up new genetic material is called transformation.

"Transformation" may also be used to describe the insertion of new genetic material into nonbacterial cells, including animal and plant cells; however, because "transformation" has a special meaning in relation to animal cells, indicating progression to a cancerous state, the term should be avoided for animal cells when describing introduction of exogenous genetic material. Introduction of foreign DNA into eukaryotic cells is often called "transfection".

History

Transformation in bacteria was first demonstrated in 1928 by British bacteriologist Frederick Griffith. Griffith discovered that a strain of *Streptococcus pneumoniae* could be made virulent after being exposed to heat-killed virulent strains. Griffith hypothesized that some "transforming principle" from the heat-killed strain was responsible for making the harmless strain virulent. In 1944 this "transforming principle" was identified as being genetic by Oswald Avery, Colin MacLeod, and Maclyn McCarty. They isolated DNA from a virulent strain of *S. pneumoniae* and using just this DNA were able to make a harmless strain virulent. They called this uptake and incorporation of DNA by bacteria "transformation" Avery-MacLeod-McCarty experiment). The results of Avery et al.'s experiments were at first skeptically received by the scientific community and it was not until the development of genetic markers and the discovery of other methods of genetic transfer (conjugation in 1947 and transduction in 1953) by Joshua Lederberg that Avery's experiments were accepted.

It was originally thought that *Escherichia coli*, a commonly used laboratory organism, was refractory to transformation. However, in 1970, Morton Mandel and Akiko Higa showed that *E. coli* may be induced to take up DNA from bacteriophage λ without the use of helper phage after treatment with calcium chloride solution. Two years later in 1972, Stanley Norman Cohen, Annie Chang and Leslie Hsu showed that CaCl 2 treatment is also effective for transformation of plasmid DNA. The method of transformation by Mandel and Higa was later improved upon by Douglas Hanahan. The discovery of artificially induced competence in *E. coli* created an efficient and convenient procedure for transforming bacteria which allows for simpler molecular cloning methods in biotechnology and research, and it is now a routinely used laboratory procedure.

Transformation using electroporation was developed in the late 1980s, increasing the efficiency of in-vitro transformation and increasing the number of bacterial strains that could be transformed. Transformation of animal and plant cells was also investigated with the first transgenic mouse being created by injecting a gene for a rat growth hormone into a mouse embryo in 1982. In 1907 a bacterium that caused plant tumors, *Agrobacterium tumefaciens*, was discovered and in the early 1970s the tumor-inducing agent was found to be a DNA plasmid called the Ti plasmid. By removing the genes in the plasmid that caused the tumor and adding in novel genes, researchers were able to infect plants with *A. tumefaciens* and let the bacteria insert their chosen DNA into the genomes of the plants. Not all plant cells are susceptible to infection by *A. tumefaciens*, so other methods were developed, including electroporation and micro-injection. Particle bombardment was made possible with the invention of the Biolistic Particle Delivery System (gene gun) by John Sanford in the 1980s.

Definitions

Bacterial transformation may be referred to as a stable genetic change brought about by the uptake of naked DNA (DNA without associated cells or proteins) to increase DNA quantity; competence refers to the state of being able to take up exogenous DNA from the environment. There are two forms of transformation and competence: natural and artificial.

Natural Competence and Transformation

About 1% of bacterial species are capable of naturally taking up DNA under laboratory conditions; more may be able to take it up in their natural environments. DNA material can be transferred between different strains of bacteria, in a process that is called horizontal gene transfer. Some species upon cell death release their DNA to be taken up by other cells, however transformation works best with DNA from closely related species. These naturally competent bacteria carry sets of genes that provide the protein machinery to bring DNA across the cell membrane(s). The transport of the exogenous

DNA into the cells may require proteins that are involved in the assembly of type IV pili and type II secretion system, as well as DNA translocase complex at the cytoplasmic membrane.

Due to the differences in structure of the cell envelope between Gram-positive and Gram-negative bacteria, there are some differences in the mechanisms of DNA uptake in these cells, however most of them share common features that involve related proteins. The DNA first binds to the surface of the competent cells on a DNA receptor, and passes through the cytoplasmic membrane via DNA translocase. Only single-stranded DNA may pass through, the other strand being degraded by nucleases in the process. The translocated single-stranded DNA may then be integrated into the bacterial chromosomes by a RecA-dependent process. In Gram-negative cells, due to the presence of an extra membrane, the DNA requires the presence of a channel formed by secretins on the outer membrane. Pilin may be required for competence, but its role is uncertain. The uptake of DNA is generally non-sequence specific, although in some species the presence of specific DNA uptake sequences may facilitate efficient DNA uptake.

Natural Transformation

Natural transformation is a bacterial adaptation for DNA transfer that depends on the expression of numerous bacterial genes whose products appear to be responsible for this process. In general, transformation is a complex, energy-requiring developmental process. In order for a bacterium to bind, take up and recombine exogenous DNA into its chromosome, it must become competent, that is, enter a special physiological state. Competence development in *Bacillus subtilis* requires expression of about 40 genes. The DNA integrated into the host chromosome is usually (but with rare exceptions) derived from another bacterium of the same species, and is thus homologous to the resident chromosome.

In *B. subtilis* the length of the transferred DNA is greater than 1271 kb (more than 1 million bases). The length transferred is likely double stranded DNA and is often more than a third of the total chromosome length of 4215 kb. It appears that about 7-9% of the recipient cells take up an entire chromosome.

The capacity for natural transformation appears to occur in a number of prokaryotes, and thus far 67 prokaryotic species (in seven different phyla) are known to undergo this process.

Competence for transformation is typically induced by high cell density and/or nutritional limitation, conditions associated with the stationary phase of bacterial growth. Transformation in *Haemophilus influenzae* occurs most efficiently at the end of exponential growth as bacterial growth approaches stationary phase. Transformation in *Streptococcus mutans*, as well as in many other streptococci, occurs at high cell density and is associated with biofilm formation. Competence in *B. subtilis* is induced toward the end of logarithmic growth, especially under conditions of amino acid limitation.

Transformation, as an Adaptation for DNA Repair

Competence is specifically induced by DNA damaging conditions. For instance, transformation is induced in *Streptococcus pneumoniae* by the DNA damaging agents mitomycin C (a DNA crosslinking agent) and fluoroquinolone (a topoisomerase inhibitor that causes double-strand breaks). In *B. subtilis*, transformation is increased by UV light, a DNA damaging agent. In *Helicobacter pylori*, ciprofloxacin, which interacts with DNA gyrase and introduces double-strand breaks, induces expression of competence genes, thus enhancing the frequency of transformation Using *Legionella pneumophila*, Charpentier et al. tested 64 toxic molecules to determine which of these induce competence. Of these only six, all DNA damaging agents, caused strong induction. These DNA damaging agents were mitomycin C (which causes DNA inter-strand crosslinks), norfloxacin, ofloxacin and nalidixic acid (inhibitors of DNA gyrase that cause double-strand breaks), bicyclomycin (causes single- and double-strand breaks), and hydroxyurea (induces DNA base oxidation). UV light also induced competence in *L. pneumophila*. Charpentier et al. suggested that competence for transformation probably evolved as a DNA damage response.

Logarithmically growing bacteria differ from stationary phase bacteria with respect to the number of genome copies present in the cell, and this has implications for the capability to carry out an important DNA repair process. During logarithmic growth, two or more copies of any particular region of the chromosome may be present in a bacterial cell, as cell division is not precisely matched with chromosome replication. The process of homologous recombinational repair (HRR) is a key DNA repair process that is especially effective for repairing double-strand damages, such as double-strand breaks. This process depends on a second homologous chromosome in addition to the damaged chromosome. During logarithmic growth, a DNA damage in one chromosome may be repaired by HRR using sequence information from the other homologous chromosome. Once cells approach stationary phase, however, they typically have just one copy of the chromosome, and HRR requires input of homologous template from outside the cell by transformation.

To test whether the adaptive function of transformation is repair of DNA damages, a series of experiments were carried out using *B. subtilis* irradiated by UV light as the damaging agent (reviewed by Michod et al. and Bernstein et al.) The results of these experiments indicated that transforming DNA acts to repair potentially lethal DNA damages introduced by UV light in the recipient DNA. The particular process responsible for repair was likely HRR. Transformation in bacteria can be viewed as a primitive sexual process, since it involves interaction of homologous DNA from two individuals to form recombinant DNA that is passed on to succeeding generations. Bacterial transformation in prokaryotes may have been the ancestral process that gave rise to meiotic sexual reproduction in eukaryotes. Evolution of sexual reproduction; Meiosis.

Methods and Mechanisms of Transformation in Laboratory

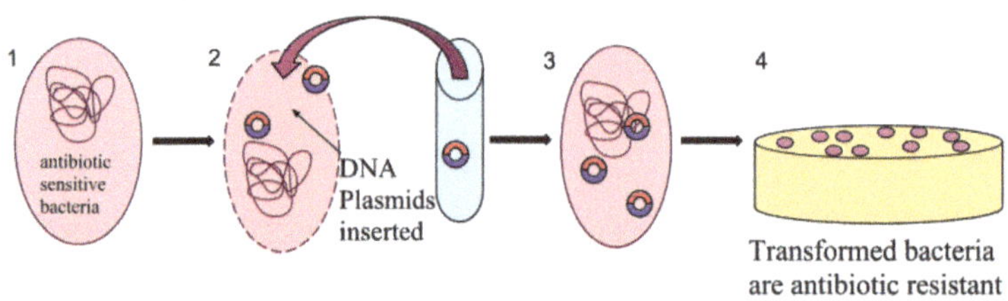

Schematic of bacterial transformation — for which artificial competence must first be induced.

Bacterial

Artificial competence can be induced in laboratory procedures that involve making the cell passively permeable to DNA by exposing it to conditions that do not normally occur in nature. Typically the cells are incubated in a solution containing divalent cations (often calcium chloride) under cold conditions, before being exposed to a heat pulse (heat shock).

It has been found that growth of Gram negative bacteria in 20 mM Mg reduces the number of protein-to-lipopolysaccharide bonds by increasing the ratio of ionic to covalent bonds, which increases membrane fluidity, facilitating transformation. The role of lipopolysaccharides here are verified from the observation that shorter O-side chains are more effectively transformed — perhaps because of improved DNA accessibility.

The surface of bacteria such as *E. coli* is negatively charged due to phospholipids and lipopolysaccharides on its cell surface, and the DNA is also negatively charged. One function of the divalent cation therefore would be to shield the charges by coordinating the phosphate groups and other negative charges, thereby allowing a DNA molecule to adhere to the cell surface.

DNA entry into *E. coli* cells is through channels known as zones of adhesion or Bayer's junction, with a typical cell carrying as many as 400 such zones. Their role was established when cobalamine (which also uses these channels) was found to competitively inhibit DNA uptake. Another type of channel implicated in DNA uptake consists of poly (HB):poly P:Ca. In this poly (HB) is envisioned to wrap around DNA (itself a polyphosphate), and is carried in a shield formed by Ca ions.

It is suggested that exposing the cells to divalent cations in cold condition may also change or weaken the cell surface structure, making it more permeable to DNA. The heat-pulse is thought to create a thermal imbalance across the cell membrane, which forces the DNA to enter the cells through either cell pores or the damaged cell wall.

Electroporation is another method of promoting competence. In this method the cells are briefly shocked with an electric field of 10-20 kV/cm, which is thought to create holes in the cell membrane through which the plasmid DNA may enter. After the electric shock, the holes are rapidly closed by the cell's membrane-repair mechanisms.

Yeast

Most species of yeast, including *Saccharomyces cerevisiae,* may be transformed by exogenous DNA in the environment. Several methods have been developed to facilitate this transformation at high frequency in the lab.

- Yeast cells may be treated with enzymes to degrade their cell walls, yielding spheroplasts. These cells are very fragile but take up foreign DNA at a high rate.
- Exposing intact yeast cells to alkali cations such as those of cesium or lithium allows the cells to take up plasmid DNA. Later protocols adapted this transformation method, using lithium acetate, polyethylene glycol, and single-stranded DNA. In these protocols, the single-stranded DNA preferentially binds to the yeast cell wall, preventing plasmid DNA from doing so and leaving it available for transformation.
- Electroporation: Formation of transient holes in the cell membranes using electric shock; this allows DNA to enter as described above for bacteria.
- Enzymatic digestion or agitation with glass beads may also be used to transform yeast cells.

Efficiency – Different yeast genera and species take up foreign DNA with different efficiencies. Also, most transformation protocols have been developed for baker's yeast, *S. cerevisiae,* and thus may not be optimal for other species. Even within one species, different strains have different transformation efficiencies, sometimes different by 3 orders of magnitude. For instance, when S. cerevisiae strains were transformed with 10 ug of plasmid YEp13, the strain DKD-5D-H yielded between 550 and 3115 colonies while strain OS1 yielded less than 5 colonies.

Plants

A number of methods are available to transfer DNA into plant cells. Some vector-mediated methods are:

- *Agrobacterium*-mediated transformation is the easiest and most simple plant transformation. Plant tissue (often leaves) are cut into small pieces, e.g. 10x10mm, and soaked for 10 minutes in a fluid containing suspended *Agro-*

bacterium. The bacteria will attach to many of the plant cells exposed by the cut. The plant cells secrete wound-related phenolic compounds which in turn act to upregulate the virulence operon of the Agrobacterium. The virulence operon includes many genes that encode for proteins that are part of a Type IV secretion system that exports from the bacterium proteins and DNA (delineated by specific recognition motifs called border sequences and excised as a single strand from the virulence plasmid) into the plant cell through a structure called a pilus. The transferred DNA (called T-DNA) is piloted to the plant cell nucleus by nuclear localization signals present in the Agrobacterium protein VirD2, which is covalently attached to the end of the T-DNA at the Right border (RB). Exactly how the T-DNA is integrated into the host plant genomic DNA is an active area of plant biology research. Assuming that a selection marker (such as an antibiotic resistance gene) was included in the T-DNA, the transformed plant tissue can be cultured on selective media to produce shoots. The shoots are then transferred to a different medium to promote root formation. Once roots begin to grow from the transgenic shoot, the plants can be transferred to soil to complete a normal life cycle (make seeds). The seeds from this first plant (called the T1, for first transgenic generation) can be planted on a selective (containing an antibiotic), or if an herbicide resistance gene was used, could alternatively be planted in soil, then later treated with herbicide to kill wildtype segregants. Some plants species, such as *Arabidopsis thaliana* can be transformed by dipping the flowers or whole plant, into a suspension of *Agrobacterium tumefaciens*, typically strain C58 (C=Cherry, 58=1958, the year in which this particular strain of *A. tumefaciens* was isolated from a cherry tree in an orchard at Cornell University in Ithaca, New York). Though many plants remain recalcitrant to transformation by this method, research is ongoing that continues to add to the list the species that have been successfully modified in this manner.

- Viral transformation (transduction): Package the desired genetic material into a suitable plant virus and allow this modified virus to infect the plant. If the genetic material is DNA, it can recombine with the chromosomes to produce transformant cells. However, genomes of most plant viruses consist of single stranded RNA which replicates in the cytoplasm of infected cell. For such genomes this method is a form of transfection and not a real transformation, since the inserted genes never reach the nucleus of the cell and do not integrate into the host genome. The progeny of the infected plants is virus-free and also free of the inserted gene.

Some Vector-less Methods Include:

- Gene gun: Also referred to as particle bombardment, microprojectile bombardment, or biolistics. Particles of gold or tungsten are coated with DNA and then shot into young plant cells or plant embryos. Some genetic material will stay in the cells and transform them. This method also allows transformation of plant plastids. The transformation efficiency is lower than in *Agrobacterium*-mediated transformation, but most plants can be transformed with this method.

- Electroporation: Formation of transient holes in cell membranes using electric pulses of high field strength; this allows DNA to enter as described above for bacteria.

Animals

Introduction of DNA into animal cells is usually called transfection, and is discussed in the corresponding article.

Practical Aspects of Transformation in Molecular Biology

The discovery of artificially induced competence in bacteria allow bacteria such as *Escherichia coli* to be used as a convenient host for the manipulation of DNA as well as expressing proteins. Typically plasmids are used for transformation in *E. coli*. In order to be stably maintained in the cell, a plasmid DNA molecule must contain an origin of replication, which allows it to be replicated in the cell independently of the replication of the cell's own chromosome.

The efficiency with which a competent culture can take up exogenous DNA and express its genes is known as transformation efficiency and is measured in colony forming unit (cfu) per µg DNA used. A transformation efficiency of 1×10^8 cfu/µg for a small plasmid like pUC19 is roughly equivalent to 1 in 2000 molecules of the plasmid used being transformed.

In calcium chloride transformation, the cells are prepared by chilling cells in the presence of Ca2+(in CaCl2 solution), making the cell become permeable to plasmid DNA. The cells are incubated on ice with the DNA, and then briefly heat-shocked (e.g., at 42 °C for 30–120 seconds). This method works very well for circular plasmid DNA. Non-commercial preparations should normally give 10^6 to 10^7 transformants per microgram of plasmid; a poor preparation will be about 10^4/µg or less, but a good preparation of competent cells can give up to ~10^8 colonies per microgram of plasmid. Protocols, however, exist for making supercompetent cells that may yield a transformation efficiency of over 10^9. The chemical method, however, usually does not work well for linear DNA, such as fragments of chromosomal DNA, probably because the cell's native exonuclease enzymes rapidly degrade linear DNA. In contrast, cells that are naturally competent are usually transformed more efficiently with linear DNA than with plasmid DNA.

The transformation efficiency using the CaCl2 method decreases with plasmid size, and electroporation therefore may be a more effective method for the uptake of large plasmid DNA. Cells used in electroporation should be prepared first by washing in cold double-distilled water to remove charged particles that may create sparks during the electroporation process.

Selection and Screening in Plasmid Transformation

Because transformation usually produces a mixture of relatively few transformed cells and an abundance of non-transformed cells, a method is necessary to select for the cells that have acquired the plasmid. The plasmid therefore requires a selectable marker such that those cells without the plasmid may be killed or have their growth arrested. Antibiotic resistance is the most commonly used marker for prokaryotes. The transforming plasmid contains a gene that confers resistance to an antibiotic that the bacteria are otherwise sensitive to. The mixture of treated cells is cultured on media that contain the antibiotic so that only transformed cells are able to grow. Another method of selection is the use of certain auxotrophic markers that can compensate for an inability to metabolise certain amino acids, nucleotides, or sugars. This method requires the use of suitably mutated strains that are deficient in the synthesis or utility of a particular biomolecule, and the transformed cells are cultured in a medium that allows only cells containing the plasmid to grow.

In a cloning experiment, a gene may be inserted into a plasmid used for transformation. However, in such experiment, not all the plasmids may contain a successfully inserted gene. Additional techniques may therefore be employed further to screen for transformed cells that contain plasmid with the insert. Reporter genes can be used as markers, such as the *lacZ* gene which codes for β-galactosidase used in blue-white screening. This method of screening relies on the principle of α-complementation, where a fragment of the *lacZ* gene (*lacZα*) in the plasmid can complement another mutant *lacZ* gene (*lacZΔM15*) in the cell. Both genes by themselves produce non-functional peptides, however, when expressed together, as when a plasmid containing *lacZ-α* is transformed into a *lacZΔM15* cells, they form a functional β-galactosidase. The presence of an active β-galactosidase may be detected when cells are grown in plates containing X-gal, forming characteristic blue colonies. However, the multiple cloning site, where a gene of interest may be ligated into the plasmid vector, is located within the *lacZα* gene. Successful ligation therefore disrupts the *lacZα* gene, and no functional β-galactosidase can form, resulting in white colonies. Cells containing successfully ligated insert can then be easily identified by its white coloration from the unsuccessful blue ones.

Other commonly used reporter genes are green fluorescent protein (GFP), which produces cells that glow green under blue light, and the enzyme luciferase, which catalyzes a reaction with luciferin to emit light. The recombinant DNA may also be detected using other methods such as nucleic acid hybridization with radioactive RNA probe, while cells that expressed the desired protein from the plasmid may also be detected using immunological methods.

Marker-assisted Selection

Marker assisted selection or marker aided selection (MAS) is an indirect selection process where a trait of interest is selected based on a marker (morphological, biochemical or DNA/RNA variation) linked to a trait of interest (e.g. productivity, disease resistance, abiotic stress tolerance, and quality), rather than on the trait itself. This process is used in plant and animal breeding.

For example, using MAS to select individuals with disease resistance involves identifying a marker allele that is linked with disease resistance rather than the level of disease resistance. The assumption is that the marker associates at high frequency with the gene or quantitative trait locus (QTL) of interest, due to genetic linkage (close proximity, on the chromosome, of the marker locus and the disease resistance-determining locus). MAS can be useful to select for traits that are difficult or expensive to measure, exhibit low heritability and/or are expressed late in development. At certain points in the breeding process the specimens are examined to ensure that they express the desired trait.

Marker Types

The majority of MAS work in the present era uses DNA-based markers. However, the first markers that allowed indirect selection of a trait of interest were morphological markers. In 1923, Sax first reported association of a simply inherited genetic marker with a quantitative trait in plants when he observed segregation of seed size associated with segregation for a seed coat color marker in beans (*Phaseolus vulgaris* L.). In 1935, Rasmusson demonstrated linkage of flowering time (a quantitative trait) in peas with a simply inherited gene for flower color.

Markers may be:

- Morphological - These markers are often detectable by eye, by simple visual inspection. Examples of this type of marker include the presence or absence of an awn, leaf sheath coloration, height, grain color, aroma of rice etc. In well-characterized crops like maize, tomato, pea, barley or wheat, tens or hundreds of genes that determine morphological traits have been mapped to specific chromosome locations.

- Biochemical- A protein that can be extracted and observed; for example, isozymes and storage proteins.

- Cytological - The chromosomal banding produced by different stains; for example, G banding.

- DNA-based or molecular- A unique gene (DNA sequence), occurring in proximity to the gene or locus of interest, can be identified by a range of molecular techniques such as RFLP, RAPD, AFLP, DAF, SCAR, microsatellite, or single-nucleotide polymorphism (SNP) detection.

Positive and Negative Selectable Markers

The following terms are generally less relevant to discussions of MAS in plant and animal breeding, but are highly relevant in molecular biology research:

- Positive selectable markers are selectable markers that confer selective advantage to the host organism. An example would be antibiotic resistance, which allows the host organism to survive antibiotic selection.
- Negative selectable markers are selectable markers that eliminate or inhibit growth of the host organism upon selection. An example would be thymidine kinase, which makes the host sensitive to ganciclovir selection.

A distinction can be made between selectable markers (which eliminate certain genotypes from the population) and screenable markers (which cause certain genotypes to be readily identifiable, at which point the experimenter must "score" or evaluate the population and act to retain the preferred genotypes). Most MAS uses screenable markers rather than selectable markers.

Gene vs Marker

The gene of interest directly causes production of protein(s) or RNA that produce a desired trait or phenotype, whereas markers (a DNA sequence or the morphological or biochemical markers produced due to that DNA) are genetically linked to the gene of interest. The gene of interest and the marker tend to move together during segregation of gametes due to their proximity on the same chromosome and concomitant reduction in recombination (chromosome crossover events) between the marker and gene of interest. For some traits, the gene of interest has been discovered and the presence of desirable alleles can be directly assayed with a high level of confidence. However, if the gene of interest is not known, markers linked to the gene of interest can still be used to select for individuals with desirable alleles of the gene of interest. When markers are used there may be some inaccurate results due to inaccurate tests for the marker. There also can be false positive results when markers are used, due to recombination between the marker of interest and gene (or QTL). A perfect marker would elicit no false positive results. The term 'perfect marker' is sometimes used when tests are performed to detect a SNP or other DNA polymorphism in the gene of interest, if that SNP or other polymorphism is the direct cause of the trait of interest. The term 'marker' is still appropriate to use when directly assaying the gene of interest, because the test of genotype is an indirect test of the trait or phenotype of interest.

Important Properties of Ideal Markers for MAS

An ideal marker:

- Easy recognition of all possible phenotypes (homo- and heterozygotes) from all different alleles

- Demonstrates measurable differences in expression between trait types or gene of interest alleles, early in the development of the organism
- Testing for the marker does not have variable success depending on the allele at the marker locus or the allele at the target locus (the gene of interest that determines the trait of interest).
- Low or null interaction among the markers allowing the use of many at the same time in a segregating population
- Abundant in number
- Polymorphic

Demerits of Morphological Markers

Morphological markers are associated with several general deficits that reduce their usefulness including:

- the delay of marker expression until late into the development of the organism
- dominance
- deleterious effects
- pleiotropy
- confounding effects of genes unrelated to the gene or trait of interest but which also affect the morphological marker (epistasis)
- rare polymorphism
- frequent confounding effects of environmental factors which affect the morphological characteristics of the organism

To avoid problems specific to morphological markers, the DNA-based markers have been developed. They are highly polymorphic, exhibit simple inheritance (often co-domimant), are abundant throughout the genome, are easy and fast to detect, exhibit minimum pleiotropic effects, and detection is not dependent on the developmental stage of the organism. Numerous markers have been mapped to different chromosomes in several crops including rice, wheat, maize, soybean and several others, and in livestock such as cattle, pigs and chickens. Those markers have been used in diversity analysis, parentage detection, DNA fingerprinting, and prediction of hybrid performance. Molecular markers are useful in indirect selection processes, enabling manual selection of individuals for further propagation.

Selection for Major Genes Linked to Markers

'Major genes' that are responsible for economically important characteristics are frequent in the plant kingdom. Such characteristics include disease resistance, male sterility, self-incompatibility, and others related to shape, color, and architecture of whole plants and are often of mono- or oligogenic in nature. The marker loci that are tightly linked to major genes can be used for selection and are sometimes more efficient than direct selection for the target gene. Such advantages in efficiency may be due for example, to higher expression of the marker mRNA in such cases that the marker is actually a gene. Alternatively, in such cases that the target gene of interest differs between two alleles by a difficult-to-detect single nucleotide polymorphism, an external marker (be it another gene or a polymorphism that is easier to detect, such as a short tandem repeat) may present as the most realistic option.

Situations that are Favorable for Molecular Marker Selection

There are several indications for the use of molecular markers in the selection of a genetic trait.

In such situations that:

- the selected character is expressed late in plant development, like fruit and flower features or adult characters with a juvenile period (so that it is not necessary to wait for the organism to become fully developed before arrangements can be made for propagation)
- the expression of the target gene is recessive (so that individuals which are heterozygous positive for the recessive allele can be crossed to produce some homozygous offspring with the desired trait)
- there is requirement for the presence of special conditions in order to invoke expression of the target gene(s), as in the case of breeding for disease and pest resistance (where inoculation with the disease or subjection to pests would otherwise be required). This advantage derives from the errors due to unreliable inoculation methods and the fact that field inoculation with the pathogen is not allowed in many areas for safety reasons. Moreover, problems in the recognition of the environmentally unstable genes can be eluded.
- the phenotype is affected by two or more unlinked genes (epistatis). For example, selection for multiple genes which provide resistance against diseases or insect pests for gene pyramiding.

The cost of genotyping (an example of a molecular marker assay) is reducing while the cost of phenotyping is increasing particularly in developed countries thus increasing the attractiveness of MAS as the development of the technology continues.

Steps for MAS

Generally the first step is to map the gene or quantitative trait locus (QTL) of interest first by using different techniques and then using this information for marker assisted selection. Generally, the markers to be used should be close to gene of interest (<5 recombination unit or cM) in order to ensure that only minor fraction of the selected individuals will be recombinants. Generally, not only a single marker but rather two markers are used in order to reduce the chances of an error due to homologous recombination. For example, if two flanking markers are used at same time with an interval between them of approximately 20cM, there is higher probability (99%) for recovery of the target gene.

QTL Mapping Techniques

In plants QTL mapping is generally achieved using bi-parental cross populations; a cross between two parents which have a contrasting phenotype for the trait of interest are developed. Commonly used populations are recombinant inbred lines (RILs), doubled haploids (DH), back cross and F_2. Linkage between the phenotype and markers which have already been mapped is tested in these populations in order to determine the position of the QTL. Such techniques are based on linkage and are therefore referred to as "linkage mapping".A

Single Step MAS and QTL Mapping

In contrast to two-step QTL mapping and MAS, a single-step method for breeding typical plant populations has been developed.

In such an approach, in the first few breeding cycles, markers linked to the trait of interest are identified by QTL mapping and later the same information is used in the same population. In this approach, pedigree structure is created from families that are created by crossing number of parents (in three-way or four way crosses). Both phenotyping and genotyping is done using molecular markers mapped the possible location of QTL of interest. This will identify markers and their favorable alleles. Once these favorable marker alleles are identified, the frequency of such alleles will be increased and response to marker assisted selection is estimated. Marker allele(s) with desirable effect will be further used in next selection cycle or other experiments.

High-throughput Genotyping Techniques

Recently high-throughput genotyping techniques are developed which allows marker aided screening of many genotypes. This will help breeders in shifting traditional breeding to marker aided selection. One example of such automation is using DNA isolation robots, capillary electrophoresis and pipetting robots.

One recent example of capllilary system is Applied Biosystems 3130 Genetic Analyzer. This is the latest generation of 4-capillary electrophoresis instruments for the low to medium throughput laboratories.

Use of MAS for Backcross Breeding

A minimum of five or six-backcross generations are required to transfer a gene of interest from a donor (may not be adapted) to a recipient (recurrent – adapted cultivar). The recovery of the recurrent genotype can be accelerated with the use of molecular markers. If the F1 is heterozygous for the marker locus, individuals with the recurrent parent allele(s) at the marker locus in first or subsequent backcross generations will also carry a chromosome tagged by the marker.

Marker Assisted Gene Pyramiding

Gene pyramiding has been proposed and applied to enhance resistance to disease and insects by selecting for two or more than two genes at a time. For example in rice such pyramids have been developed against bacterial blight and blast. The advantage of use of markers in this case allows to select for QTL-allele-linked markers that have same phenotypic effect.

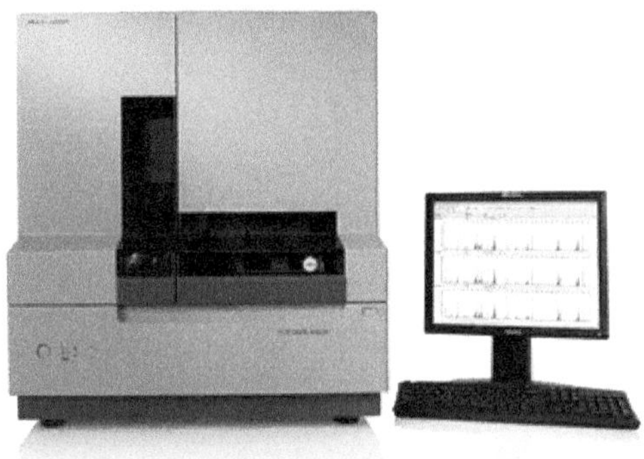

ABI 3130 genetic analyzer.

MAS has also been proved useful for livestock improvement.

A coordinated effort to implement wheat (*Triticum turgidum* and *Triticum aestivum)* marker assisted selection in the U.S. as well as a resource for marker assisted selection exists at the Wheat CAP (Coordinated Agricultural Project) website. Farhad Kahani

Marker Assisted Selection Learning Lessons

- Using OptiMAS for Marker-Assisted Recurrent Selection in a Multi-Parental Population
- Video Overview of DNA Extraction and Marker-Assisted Selection
- Marker Assisted Selection
- Background Selection

- Forward Selection
- Marker Assisted Selection in Tomato
- Marker-Assisted Selection for Bacterial Spot Resistance in Tomato Case Study
- Marker-Assisted Selection for PVY Resistance in Potato
- Marker-Assisted Selection for Resistance to Golden Nematode in Potato
- Gene Pyramiding Using Molecular Markers
- Pyramiding Resistance Genes for Bacterial Spot and Bacterial Speck on Tomato Chromosome 5
- The SolCAP Tomato Phenotypic Data: Estimating Heritability and BLUPs for Traits
- rrBLUP Package in R for Genomewide Selection

Various Types of Markers

Morphology (Biology)

Morphology of a male *Caprella mutica*

Morphology is a branch of biology dealing with the study of the form and structure of organisms and their specific structural features.

This includes aspects of the outward appearance (shape, structure, colour, pattern, size), i.e. external morphology (or eidonomy), as well as the form and structure of the internal parts like bones and organs, i.e. internal morphology (or anatomy). This is in contrast to physiology, which deals primarily with function. Morphology is a branch of life science dealing with the study of gross structure of an organism or taxon and its component parts.

Etymology and Term Usage

The word "morphology" is from the Ancient Greek *morphé*, meaning "form", and *lógos*, meaning "word, study, research". The biological concept of morphology was developed by Johann Wolfgang von Goethe (1790) and independently by the German anatomist and physiologist Karl Friedrich Burdach (1800).

Among other important theorists of morphology are Oken, Cuvier, Geoffroy, Owen, Gegenbaur and Haeckel.

In English-speaking countries, the term "molecular morphology" has been used for some time for describing the structure of compound molecules, such as polymers and RNA. The term "gross morphology" refers to the collective structures or an organism as a whole as a general description of the form and structure of an organism, taking into account all of its structures without specifying an individual structure.

Branches of Morphology

- Comparative Morphology is analysis of the patterns of the locus of structures within the body plan of an organism, and forms the basis of taxonomical categorization.
- Functional Morphology is the study of the relationship between the structure and function of morphological features.
- Experimental Morphology is the study of the effects of external factors upon the morphology of organisms under experimental conditions, such as the effect of genetic mutation.
- "Anatomy" is a "branch of morphology that deals with the structure of organisms".

Morphology and Classification

Most taxa differ morphologically from other taxa. Typically, closely related taxa differ much less than more distantly related ones, but there are exceptions to this. Cryptic species are species which look very similar, or perhaps even outwardly identical, but are reproductively isolated. Conversely, sometimes unrelated taxa acquire a similar appearance as a result of convergent evolution or even mimicry. In addition, there can be morphological differences within a species, such as in *Apoica flavissima* where queens are significantly smaller than workers. A further problem with relying on morphological data is that what may appear, morphologically speaking, to be two distinct species, may in fact be shown by DNA analysis to be a single species. The significance of these differences can be examined through the use of allometric engineering in which one or both species are manipulated to phenocopy the other species.

3D Cell Morphology:classification

Invention and development of microscopy enable the observation of 3-D cell morphology with both high spatial and temporal resolution. The dynamic processes of these cell morphology which are controlled by a complex system play an important role in varied important biological process, such as immune and invasive responses.

Cell Biology

Understanding the cell in terms of its generalized structure and molecular components.

Cell biology is a branch of biology that studies the different structures and functions of the cell and focuses mainly on the idea of the cell as the basic unit of life. Cell biology explains the structure, organization of the organelles they contain, their physiological properties, metabolic processes, signaling pathways, life cycle, and interactions with their environment. This is done both on a microscopic and molecular level as it encompasses prokaryotic cells and eukaryotic cells. Knowing the components of cells and how cells work is fundamental to all biological sciences; it is also essential for research in bio-medical fields such as cancer, and other diseases. Research in cell biology is closely related to genetics, biochemistry, molecular biology, immunology, and developmental biology.

Internal Cellular Structures

Chemical and Molecular Environment

The study of the cell is done on a molecular level; however, most of the processes within the cell are made up of a mixture of small organic molecules, inorganic ions, hormones, and water. Approximately 75-85% of the cell's volume is due to water making it an indispensable solvent as a result of its polarity and structure. These molecules within the cell, which operate as substrates, provide a suitable environment for the cell to carry out metabolic reactions and signalling. The cell shape varies among the different types of organisms, and are thus then classified into two categories: eukaryotes and prokaryotes. In the case of eukaryotic cells - which are made up of animal, plant, fungi, and protozoa cells - the shapes are generally round and spherical, while for prokaryotic cells – which are composed of bacteria and archaea - the shapes are: spherical (cocci), rods (bacillus), curved (vibrio), and spirals (*spirochetes).*

Cell biology focuses more on the study of eukaryotic cells, and their signalling pathways, rather than on prokaryotes which is covered under microbiology. The main constituents of the general molecular composition of the cell includes: proteins and lipids which are either free flowing or membrane bound, along with different internal compartments known as organelles. This environment of the cell is made up of hydrophilic and hydrophobic regions which allows for the exchange of the above-mentioned molecules and ions. The hydrophilic regions of the cell are mainly on the inside and outside of the cell, while the hydrophobic regions are within the phospholipid bilayer of the cell membrane. The cell membrane consists of lipids and proteins which accounts for its hydrophobicity as a result of being non-polar substances. Therefore, in order for these molecules to participate in reactions, within the cell, they need to be able to cross this membrane layer to get into the cell. They accomplish this process of gaining access to the cell via: osmotic pressure, diffusion, concentration gradients, and membrane channels. Inside of the cell are extensive internal sub-cellular membrane-bounded compartments called organelles.

Organelles

- Centrosome - an associated pair of cylindrical shaped protein structures (centrioles) that organize microtubules and aid in forming the mitotic spindle during cell division in eukaryotes
- Cell membrane (plasma membrane) - the part of the cell which separates the cells from the outside environment and protects the cell, as well as regulating what goes in and out of the cell
- Cell wall - extra layer of protection and gives structural support (only found in plant cells)
- Chloroplast - key organelle for photosynthesis (only found in plant cells)
- Cilium - motile structure of eukaryotes having a cytoskeleton, the axoneme.
- Cytoplasm - contents of the main fluid-filled space inside cells, chemical reactions also happen in this jelly-like substance.
- Cytoskeleton - protein filaments inside cells (microfilaments, microtubules, and intermediate filaments)
- Endoplasmic reticulum (rough) - major site of membrane protein synthesis
- Endoplasmic reticulum (smooth) - major site of lipid synthesis
- Endosomes - vesicles that traffic membrane and intra and extra cellular contents for recycling or degradation by lysosomes

- Flagellum - motile structure of bacteria, archaea and eukaryotes
- Golgi apparatus - site of protein glycosylation in the endomembrane system
- Lipid bilayer - fundamental organizational structure of cell membranes
- Lysosome - acidic organelle that breaks down cellular waste products and debris into simple compounds (only found in animal cells)
- Microvilli - increases surface area for absorption of nutrients from surrounding medium
- Mitochondrion - major energy-producing organelle by releasing energy in the form of ATP
- Nucleus - contains chromosomes composed of DNA, the building block of life. Nuclear Architecture is important for dictating nuclear function.
- Organelle - term used for major subcellular structures
- Peroxisomes - a very small organelle that uses oxygen to breakdown and detoxify long fatty acids and other molecules
- Pili - also called fimbria is used for conjugation and sometimes movement
- Ribosome - RNA and protein complex required for protein synthesis in cells
- Starch grain - found in the cytoplasm of a typical plant cell, it stores chemical energy of the plant.
- Vacuole - contain cell sap (only found in plant cells)
- Vesicle - small membrane-bounded spheres inside cells

cell surface membrane protects the cell

Processes

Growth and Development

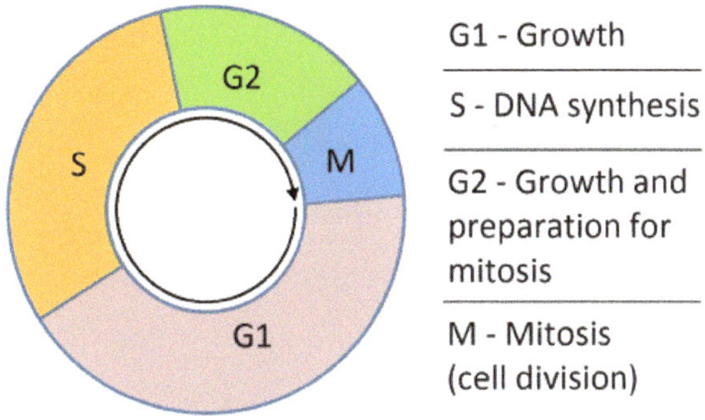

General concept of the cell cycle.

The growth process of the cell does not refer to the size of the cell, but instead the density of the number of cells present in the organism at a given time. Cell growth pertains to the increase in the number of cells present in an organism as it grows and develops; as the organism gets larger so too does the number of cells present. Cells are the foundation of all organisms, they are the fundamental unit of life. The growth and development of the cell are essential for the maintenance of the host, and survival of the organisms. For this process the cell goes through the steps of the cell cycle and development which involves cell growth, DNA replication, cell division, regeneration, specialization, and cell death. The cell cycle is divided into four distinct phases, G1, S, G2, and M. The G phases – which is the cell growth phase - makes up approximately 95% of the cycle. The proliferation of cells is instigated by progenitors, the cells then differentiate to become specialized, where specialized cells of the same type aggregate to form tissues, then organs and ultimately systems. The G phases along with the S phase – DNA replication, damage and repair - are considered to be the interphase portion of the cycle. While the M phase (mitosis and cytokinesis) is the cell division portion of the cycle. The cell cycle is regulated by a series of signalling factors and complexes such as CDK's, kinases, and p53. to name a few. When the cell has completed its growth process, and if it is found to be damaged or altered it undergoes cell death, either by apoptosis or necrosis, to eliminate the threat it cause to the organism's survival.

Other Cellular Processes

- Active transport and Passive transport - Movement of molecules into and out of cells.
- Autophagy - The process whereby cells "eat" their own internal components or microbial invaders.
- Adhesion - Holding together cells and tissues.
- Cell movement - Chemotaxis, contraction, cilia and flagella.
- Cell signaling - Regulation of cell behavior by signals from outside.
- Division - By which cells reproduce either by mitosis (to produce clones of the parent cell) or Meiosis (to produce haploid gametes)
- DNA repair - Cell death and cell senescence.
- Metabolism - Glycolysis, respiration, photosynthesis, and chemosynthesis.
- Signalling - The process by which the activities in the cell are regulated
- Transcription and mRNA splicing - Gene expression.

Techniques used to Study Cells

Cells may be observed under the microscope, using several different techniques; these include optical microscopy, transmission electron microscopy, scanning electron microscopy, fluorescence microscopy, and confocal microscopy.

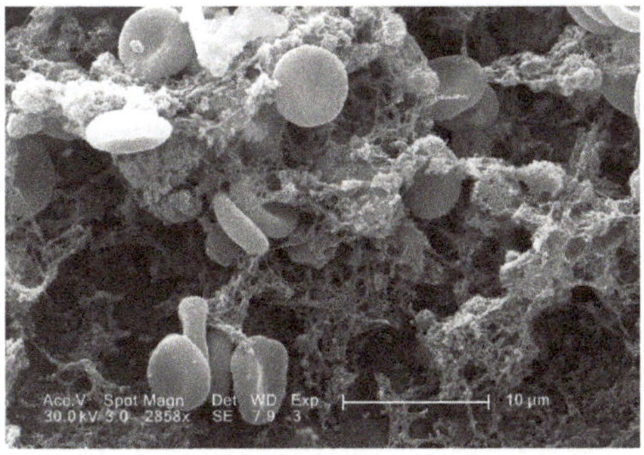

Electron micrograph of blood cells clotting.

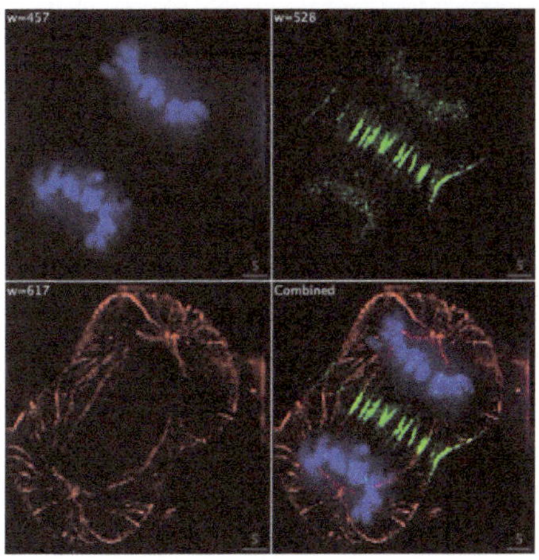

Cell division studied using fluorescence to stain specific structures

There are several different methods used in the study of cells:

- Cell culture is the basic technique of growing cells in a laboratory independent of an organism.
- Immunostaining, also known as immunohistochemistry, is a specialized histological method used to localize proteins in cells or tissue slices. Unlike regular histology, which uses stains to identify cells, cellular components or protein classes, immunostaining requires the reaction of an antibody directed against the protein of interest within the tissue or cell. Through the use of proper controls and published protocols (need to add reference links here), specificity of the antibody-antigen reaction can be achieved. Once this complex is formed, it is identified via either a "tag" attached directly to the antibody, or added in an additional technical step. Commonly used "tags" include fluorophores or enzymes. In the case of the former, detection of the location of the "immuno-stained" protein occurs via fluorescence microscopy. With an enzymatic tag, such as horse radish peroxidase, a chemical reaction is carried out that results

in a dark color in the location of the protein of interest. This darkened pattern is then detected using light microscopy.

- Computational genomics is used to find patterns in genomic information
- DNA microarrays identify changes in transcript levels between different experimental conditions.
- Gene knockdown mutates a selected gene.
- In situ hybridization shows which cells are expressing a particular RNA transcript.
- PCR can be used to determine how many copies of a gene are present in a cell.
- Transfection introduces a new gene into a cell, usually an expression construct

Purification of cells and their parts Purification may be performed using the following methods:

- Cell fractionation
  - Release of cellular organelles by disruption of cells.
  - Separation of different organelles by centrifugation.
- Flow cytometry
- Immunoprecipitation
  - The binding of an antibody to a target protein
  - Collection of the target protein through elution
- Proteins extracted from cell membranes by detergents and salts or other kinds of chemicals.

Doubled Haploidy

A doubled haploid (DH) is a genotype formed when haploid cells undergo chromosome doubling. Artificial production of doubled haploids is important in plant breeding.

Haploid cells are produced from pollen or egg cells or from other cells of the gametophyte, then by induced or spontaneous chromosome doubling, a doubled haploid cell is produced, which can be grown into a doubled haploid plant. If the original plant was diploid, the haploid cells are monoploid, and the term doubled monoploid may be used for the doubled haploids. Haploid organisms derived from tetraploids or hexaploids are sometimes called dihaploids (and the doubled dihaploids are, respectively, tetraploid or hexaploid).

Conventional inbreeding procedures take six generations to achieve approximately complete homozygosity, whereas doubled haploidy achieves it in one generation. Dihaploid plants derived from tetraploid crop plants may be important for breeding programs that involve diploid wild relatives of the crops.

History

The first report of the haploid plant was published by Blakeslee *et al.* (1922) in *Datura stramonium*. Subsequently, haploids were reported in many other species. Guha and Maheshwari (1964) developed an anther culture technique for the production of haploids in the laboratory. Haploid production by wide crossing was reported in barley (Kasha and Kao, 1970) and tobacco (Burk *et al.*, 1979). Tobacco, rapeseed, and barley are the most responsive species for doubled haploid production. Doubled haploid methodologies have now been applied to over 250 species.

Production of Doubled Haploids

Doubled haploids can be produced *in vivo* or *in vitro*. Haploid embryos are produced *in vivo* by parthenogenesis, pseudogamy, or chromosome elimination after wide crossing. The haploid embryo is rescued, cultured, and chromosome-doubling produces doubled haploids. The *in vitro* methods include gynogenesis (ovary and flower culture) and androgenesis (anther and microspore culture). Androgenesis is preferred method. Another method of producing the haploids is wide crossing. In barley, haploids can be produced by wide crossing with the related species *Hordeum bulbosum*; fertilization is affected, but during the early stages of seed development the *H. bulbosum* chromosomes are eliminated leaving a haploid embryo. In tobacco (*Nicotiana tabacum*), wide crossing with *Nicotiana africana* is widely used. When *N. africana* is used to pollinate *N. tabacum*, 0.25 to 1.42 percent of the progeny survive and can readily be identified as either F1 hybrids or maternal haploids. Although these percentages appear small, the vast yield of tiny seeds and the early death of most seedlings provide significant numbers of viable hybrids and haploids in relatively small soil containers. This method of interspecific pollination serves as a practical way of producing seed-derived haploids of *N. tabacum*, either as an alternative method or complementary method to anther culture.

Genetics of DH Population

In DH method only two types of genotypes occur for a pair of alleles, A and a, with the frequency of ½ AA and ½ aa, while in diploid method three genotypes occur with the frequency of ¼ AA, ½ Aa, ¼ aa. Thus, if AA is desirable genotype, the probability of obtaining this genotype is higher in haploid method than in diploid method. If n loci are segregating, the probability of getting the desirable genotype is (1/2)n by the haploid method and (1/4)n by the diploid method. Hence the efficiency of the haploid method is high when the number of genes concerned is large.

Studies were conducted comparing DH method and other conventional breeding methods and it was concluded that adoption of doubled haploidy does not lead to any bias of genotypes in populations, and random DHs were even found to be compatible to selected line produced by conventional pedigree method.

Applications of DHs Plant Breeding

Mapping Quantitative Trait Loci

Most of the economic traits are controlled by genes with small but cumulative effects. Although the potential of DH populations in quantitative genetics has been understood for some time, it was the advent of molecular marker maps that provided the impetus for their use in identifying loci controlling quantitative traits. As the quantitative trait loci (QTL) effects are small and highly influenced by environmental factors, accurate phenotyping with replicated trials is needed. This is possible with doubled haploidy organisms because of their true breeding nature and because they can conveniently be produced in large numbers. Using DH populations, 130 quantitative traits have been mapped in nine crop species. In total, 56 DH populations were used for QTL detection.

Backcross Breeding

In backcross conversion, genes are introgressed from a donor cultivar or related species into a recipient elite line through repeated backcrossing. A problem in this procedure is being able to identify the lines carrying the trait of interest at each generation. The problem is particularly acute if the trait of interest is recessive, as it will be present only in a heterozygous condition after each backcross. The development of molecular markers provides an easier method of selection based on the genotype (marker) rather than the phenotype. Combined with doubled haploidy it becomes more effective. In marker assisted backcross conversion, a recipient parent is crossed with a donor line and the hybrid (F1) backcrossed to the recipient. The resulting generation (BC1) is backcrossed and the process repeated until the desired genotypes are produced. The combination of doubled haploidy and molecular marker provides the short cut. In the back cross generation one itself a genotype with the character of interest can be selected and converted into homozygous doubled haploid genotype. Chen *et al.* (1994) used marker assisted backcross conversion with doubled haploidy of BC1 individuals to select stripe rust resistant lines in barley.

Bulked Segregant Analysis (BSA)

In bulked segregant analysis, a population is screened for a trait of interest and the genotypes at the two extreme ends form two bulks. Then the two bulks are tested for the presence or absence of molecular markers. Since the bulks are supposed to contrast in the alleles that contribute positive and negative effects, any marker polymorphism between the two bulks indicates the linkage between the marker and trait of interest. BSA is dependent on accurate phenotyping and the DH population has particular advantage in that they are true breeding and can be tested repeatedly. DH populations are commonly used in bulked segregant analysis, which is a popular method in marker assisted breeding. This method has been applied mostly to rapeseed and barley.

Genetic Maps

Genetic maps are very important to understand the structure and organization of genomes from which evolution patterns and syntenic relationships between species can be deduced. Genetic maps also provide a framework for the mapping of genes of interest and estimating the magnitude of their effects and aid our understanding of genotype/phenotype associations. DH populations have become standard resources in genetic mapping for species in which DHs are readily available. Doubled haploid populations are ideal for genetic mapping. It is possible to produce a genetic map within two years of the initial cross regardless of the species. Map construction is relatively easy using a DH population derived from a hybrid of two homozygous parents as the expected segregation ratio is simple, *i.e.* 1:1. DH populations have now been used to produce genetic maps of barley, rapeseed, rice, wheat, and pepper. DH populations played a major role in facilitating the generation of the molecular marker maps in eight crop species.

Genetic Studies

Genetic ratios and mutation rates can be read directly from haploid populations. A small doubled haploid (DH) population was used to demonstrate that a dwarfing gene in barley is located chromosome 5H. In another study the segregation of a range of markers has been analyzed in barley.

Genomics

Although QTL analysis has generated a vast amount of information on gene locations and the magnitude of effects on many traits, the identification of the genes involved has remained elusive. This is due to poor resolution of QTL analysis. The solution for this problem would be production of recombinant chromosome substitution line, or stepped aligned recombinant inbred lines. Here, backcrossing is carried out until a desired level of recombination has occurred and genetic markers are used to detect desired recombinant chromosome substitution lines in the target region, which can be fixed by doubled haploidy. In rice, molecular markers have been found to be linked with major genes and QTLs for resistance to rice blast, bacterial blight, and sheath blight in a map produced from DH population.

Elite Crossing

Traditional breeding methods are slow and take 10–15 years for cultivar development. Another disadvantage is inefficiency of selection in early generations because of heterozygosity. These two disadvantages can be over come by DHs, and more elite crosses can be evaluated and selected within less time.

Cultivar Development

Uniformity is a general requirement of cultivated line in most species, which can be easily obtained through DH production. There are various ways in which DHs can be used in cultivar production. The DH lines themselves can be released as cultivars, they may be used as parents in hybrid cultivar production or more indirectly in the creation of breeders lines and in germplasm conservation. Barley has over 100 direct DH cultivars. According to published information there are currently around 300 DH derived cultivars in 12 species worldwide.

The relevance of DHs to plant breeding has increased markedly in recent years owing to the development of protocols for 25 species. Doubled haploidy already plays an important role in hybrid cultivar production of vegetables, and the potential for ornamental production is being vigorously examined. DHs are also being developed in the medicinal herb *Valeriana officinalis* to select lines with high pharmacological activity. Another interesting development is that fertile homozygous DH lines can be produced in species that have self-incompatibility systems.

Advantages of DHs

The ability to produce homozygous lines after a single round recombination saves a lot of time for the plant breeders. Studies conclude that random DH's are comparable to the selected lines in pedigree inbreeding. The other advantages include development of large number of homozygous lines, efficient genetic analysis and development of markers for useful traits in much less time. More specific benefits include the possibility of seed propagation as an alternative to vegetative multiplication in ornamentals, and in species such as trees in which long life cycles and inbreeding depression preclude traditional breeding methods, doubled haploidy provides new alternatives.

Disadvantages of DHs

The main disadvantage with the DH population is that selection cannot be imposed on the population. But in conventional breeding selection can be practised for several generations: thereby desirable characters can be improved in the population.

In haploids produced from anther culture, it is observed that some plants are aneuploids and some are mixed haploid-diploid types. Another disadvantage associated with the double haploidy is the cost involved in establishing tissue culture and growth facilities. The over-usage of doubled haploidy may reduce genetic variation in breeding germplasm. Hence one has to take several factors into consideration before deploying doubled haploidy in breeding programmes.

Conclusions

Technological advances have now provided DH protocols for most plant genera. The number of species amenable to doubled haploidy has reached a staggering 250 in just a few decades. Response efficiency has also improved with gradual removal of species from recalcitrant category. Hence it will provide greater efficiency of plant breeding.

Genetic Engineering

Genetic engineering, also called genetic modification, is the direct manipulation of an organism's genome using biotechnology. It is a set of technologies used to change the genetic makeup of cells, including the transfer of genes within and across species boundaries to produce improved or novel organisms. New DNA may be inserted in the host genome by first isolating and copying the genetic material of interest using molecular cloning methods to generate a DNA sequence, or by synthesizing the DNA, and then inserting this construct into the host organism. Genes may be removed, or "knocked out", using a nuclease. Gene targeting is a different technique that uses homologous recombination to change an endogenous gene, and can be used to delete a gene, remove exons, add a gene, or introduce point mutations.

An organism that is generated through genetic engineering is considered to be a genetically modified organism (GMO). The first GMOs were bacteria generated in 1973 and GM mice in 1974. Insulin-producing bacteria were commercialized in 1982 and genetically modified food has been sold since 1994. GloFish, the first GMO designed as a pet, was first sold in the United States in December 2003.

Genetic engineering techniques have been applied in numerous fields including research, agriculture, industrial biotechnology, and medicine. Enzymes used in laundry detergent and medicines such as insulin and human growth hormone are now manufactured in GM cells, experimental GM cell lines and GM animals such as mice or zebrafish are being used for research purposes, and genetically modified crops have been commercialized.

IUPAC Definition

Process of inserting new genetic information into existing cells in order to modify a specific organism for the purpose of changing its characteristics.

Definition

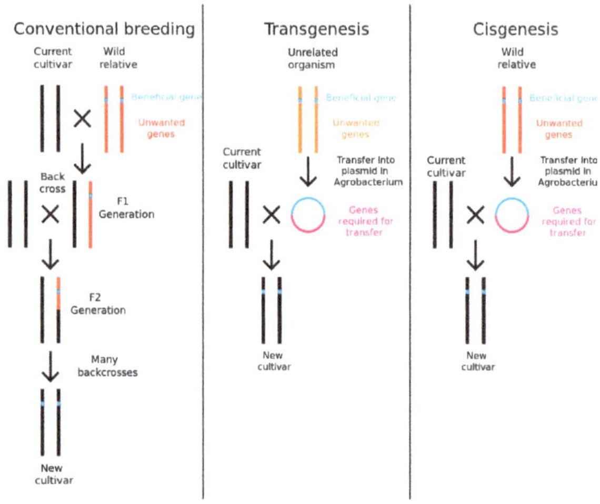

Comparison of conventional plant breeding with transgenic and cisgenic genetic modification.

Genetic engineering alters the genetic make-up of an organism using techniques that remove heritable material or that introduce DNA prepared outside the organism either directly into the host or into a cell that is then fused or hybridized with the host. This involves using recombinant nucleic acid (DNA or RNA) techniques to form new combinations of heritable genetic material followed by the incorporation of that material either indirectly through a vector system or directly through micro-injection, macro-injection and micro-encapsulation techniques.

Genetic engineering does not normally include traditional animal and plant breeding, in vitro fertilisation, induction of polyploidy, mutagenesis and cell fusion techniques that do not use recombinant nucleic acids or a genetically modified organism in the process. However the European Commission has also defined genetic engineering broadly as including selective breeding and other means of artificial selection. Cloning and stem cell research, although not considered genetic engineering, are closely related and genetic engineering can be used within them. Synthetic biology is an emerging discipline that takes genetic engineering a step further by introducing artificially synthesized material from raw materials into an organism.

If genetic material from another species is added to the host, the resulting organism is called transgenic. If genetic material from the same species or a species that can naturally breed with the host is used the resulting organism is called cisgenic. Genetic engineering can also be used to remove genetic material from the target organism, creating a gene knockout organism. In Europe genetic modification is synonymous with genetic engineering while within the United States of America it can also refer to conventional breeding methods. The Canadian regulatory system is based on whether a product has novel features regardless of method of origin. In other words, a product is regulated as genetically modified if it carries some trait not previously found in the species whether it was generated using traditional breeding methods (e.g., selective breeding, cell fusion, mutation breeding) or genetic engineering. Within the scientific community, the term *genetic engineering* is not commonly used; more specific terms such as *transgenic* are preferred.

Genetically Modified Organisms

Plants, animals or micro organisms that have changed through genetic engineering are termed genetically modified organisms or GMOs. Bacteria were the first organisms to be genetically modified. Plasmid DNA containing new genes can be inserted into the bacterial cell and the bacteria will then express those genes. These genes can code for medicines or enzymes that process food and other substrates. Plants have been modified for insect protection, herbicide resistance, virus resistance, enhanced nutrition, tolerance to environmental pressures and the production of edible vaccines. Most commercialised GMO's are insect resistant and/or herbicide tolerant crop plants. Genetically modified animals have been used for research, model animals and the production of agricultural or pharmaceutical products.

The genetically modified animals include animals with genes knocked out, increased susceptibility to disease, hormones for extra growth and the ability to express proteins in their milk.

History

Humans have altered the genomes of species for thousands of years through selective breeding, or artificial selection as contrasted with natural selection, and more recently through mutagenesis. Genetic engineering as the direct manipulation of DNA by humans outside breeding and mutations has only existed since the 1970s. The term "genetic engineering" was first coined by Jack Williamson in his science fiction novel *Dragon's Island*, published in 1951 – one year before DNA's role in heredity was confirmed by Alfred Hershey and Martha Chase, and two years before James Watson and Francis Crick showed that the DNA molecule has a double-helix structure – though the general concept of direct genetic manipulation was explored in rudimentary form in Stanley G. Weinbaum's 1936 science fiction story *Proteus Island.*

In 1972, Paul Berg created the first recombinant DNA molecules by combining DNA from the monkey virus SV40 with that of the lambda virus. In 1973 Herbert Boyer and Stanley Cohen created the first transgenic organism by inserting antibiotic resistance genes into the plasmid of an *E. coli* bacterium. A year later Rudolf Jaenisch created a transgenic mouse by introducing foreign DNA into its embryo, making it the world's first transgenic animal. These achievements led to concerns in the scientific community about potential risks from genetic engineering, which were first discussed in depth at the Asilomar Conference in 1975. One of the main recommendations from this meeting was that government oversight of recombinant DNA research should be established until the technology was deemed safe.

In 1974 Rudolf Jaenisch created the first GM animal.

In 1976 Genentech, the first genetic engineering company, was founded by Herbert Boyer and Robert Swanson and a year later the company produced a human protein (somatostatin) in *E.coli*. Genentech announced the production of genetically engineered human insulin in 1978. In 1980, the U.S. Supreme Court in the *Diamond v. Chakrabarty* case ruled that genetically altered life could be patented. The insulin produced by bacteria, branded humulin, was approved for release by the Food and Drug Administration in 1982.

In the 1970s graduate student Steven Lindow of the University of Wisconsin–Madison with D.C. Arny and C. Upper found a bacterium he identified as *P. syringae* that played a role in ice nucleation, and in 1977 he discovered a mutant ice-minus strain. Dr. Lindow (who is now a plant pathologist at the University of California-Berkeley) later successfully created a recombinant ice-minus strain. In 1983, a biotech company, Advanced Genetic Sciences (AGS) applied for U.S. government authorization to perform field tests with the ice-minus strain of *P. syringae* to protect crops from frost, but environmental groups and protestors delayed the field tests for four years with legal challenges. In 1987, the ice-minus strain of *P. syringae* became the first genetically modified organism (GMO) to be released into the environment when a strawberry field and a potato field in California were sprayed with it. Both test fields were attacked by activist groups the night before the tests occurred: "The world's first trial site attracted the world's first field trasher".

The first field trials of genetically engineered plants occurred in France and the USA in 1986, tobacco plants were engineered to be resistant to herbicides. The People's Republic of China was the first country to commercialize transgenic plants, introducing a virus-resistant tobacco in 1992. In 1994 Calgene attained approval to commercially release the Flavr Savr tomato, a tomato engineered to have a longer shelf life. In 1994, the European Union approved tobacco engineered to be resistant to the herbicide bromoxynil, making it the first genetically engineered crop commercialized in Europe. In 1995, Bt Potato was approved safe by the Environmental Protection Agency, after having been approved by the FDA, making it the first pesticide producing crop to be approved in the USA. In 2009 11 transgenic crops were grown commercially in 25 countries, the largest of which by area grown were the USA, Brazil, Argentina, India, Canada, China, Paraguay and South Africa.

In 2010, scientists at the J. Craig Venter Institute created the first synthetic genome and inserted it into an empty bacterial cell. The resulting bacterium, named Synthia, could replicate and produce proteins. In 2014, a bacterium was developed that replicated a plasmid containing a unique base pair, creating the first organism engineered to use an expanded genetic alphabet.

Process

The first step is to choose and isolate the gene that will be inserted into the genetically modified organism. The gene can be isolated using restriction enzymes to cut DNA into fragments and gel electrophoresis to separate them out according to length. Polymerase chain reaction (PCR) can also be used to amplify up a gene segment, which can then be isolated through gel electrophoresis. If the chosen gene or the donor organism's genome has been well studied it may be present in a genetic library. If the DNA sequence is known, but no copies of the gene are available, it can be artificially synthesized.

The gene to be inserted into the genetically modified organism must be combined with other genetic elements in order for it to work properly. The gene can also be modified at this stage for better expression or effectiveness. As well as the gene to be inserted most constructs contain a promoter and terminator region as well as a selectable marker gene. The promoter region initiates transcription of the gene and can be used to control the location and level of gene expression, while the terminator region ends transcription. The selectable marker, which in most cases confers antibiotic resistance to the organism it is expressed in, is needed to determine which cells are transformed with the new gene. The constructs are made using recombinant DNA techniques, such as restriction digests, ligations and molecular cloning. The manipulation of the DNA generally occurs within a plasmid.

The most common form of genetic engineering involves inserting new genetic material randomly within the host genome. Other techniques allow new genetic material to be inserted at a specific location in the host genome or generate mutations at desired genomic loci capable of knocking out endogenous genes. The technique of gene targeting uses homologous recombination to target desired changes to a specific endogenous gene. This tends to occur at a relatively low frequency in plants and animals and generally requires the use of selectable markers. The frequency of gene targeting can be greatly enhanced with the use of engineered nucleases such as zinc finger nucleases, engineered homing endonucleases, or nucleases created from TAL effectors.

In addition to enhancing gene targeting, engineered nucleases can also be used to introduce mutations at endogenous genes that generate a gene knockout.

Transformation

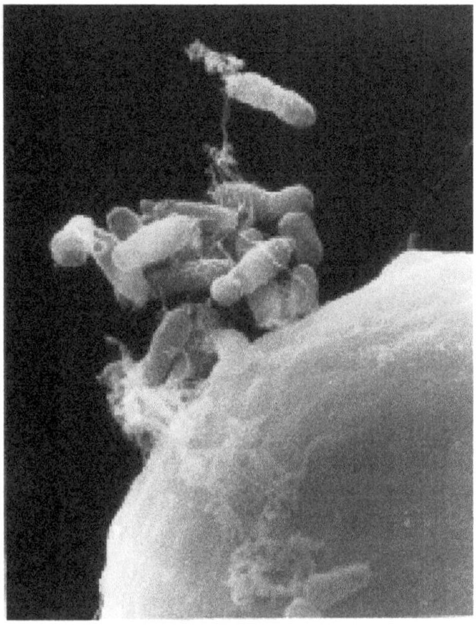

A. tumefaciens attaching itself to a carrot cell

Only about 1% of bacteria are naturally capable of taking up foreign DNA. However, this ability can be induced in other bacteria via stress (e.g. thermal or electric shock), thereby increasing the cell membrane's permeability to DNA; up-taken DNA can either integrate with the genome or exist as extrachromosomal DNA. DNA is generally inserted into animal cells using microinjection, where it can be injected through the cell's nuclear envelope directly into the nucleus or through the use of viral vectors. In plants the DNA is generally inserted using *Agrobacterium*-mediated recombination or biolistics.

In *Agrobacterium*-mediated recombination, the plasmid construct contains T-DNA, DNA which is responsible for insertion of the DNA into the host plants genome. This plasmid is transformed into *Agrobacterium* containing no plasmids prior to infecting the plant cells. The *Agrobacterium* will then naturally insert the genetic material into the plant cells. In biolistics transformation particles of gold or tungsten are coated with DNA and then shot into young plant cells or plant embryos. Some genetic material will enter the cells and transform them. This method can be used on plants that are not susceptible to *Agrobacterium* infection and also allows transformation of plant plastids. Another transformation method for plant and animal cells is electroporation. Electroporation involves subjecting the plant or animal cell to an electric shock, which can make the cell membrane permeable to plasmid DNA. In some cases the electroporated cells will incorporate the DNA into their genome. Due to the damage caused to the cells and DNA the transformation efficiency of biolistics and electroporation is lower than agrobacterial mediated transformation and microinjection.

As often only a single cell is transformed with genetic material the organism must be regenerated from that single cell. As bacteria consist of a single cell and reproduce clonally regeneration is not necessary. In plants this is accomplished through the use of tissue culture. Each plant species has different requirements for successful regeneration through tissue culture. If successful an adult plant is produced that contains the transgene in every cell. In animals it is necessary to ensure that the inserted DNA is present in the embryonic stem cells. Selectable markers are used to easily differentiate transformed from untransformed cells. These markers are usually present in the transgenic organism, although a number of strategies have been developed that can remove the selectable marker from the mature transgenic plant. When the offspring is produced they can be screened for the presence of the gene. All offspring from the first generation will be heterozygous for the inserted gene and must be mated together to produce a homozygous animal.

Further testing uses PCR, Southern hybridization, and DNA sequencing is conducted to confirm that an organism contains the new gene. These tests can also confirm the chromosomal location and copy number of the inserted gene. The presence of the gene does not guarantee it will be expressed at appropriate levels in the target tissue so methods that look for and measure the gene products (RNA and protein) are also used. These include northern hybridization, quantitative RT-PCR, Western blot, immunofluorescence, ELISA and phenotypic analysis. For stable transformation the gene should be passed to the offspring in a Mendelian inheritance pattern, so the organism's offspring are also studied.

Genome Editing

Genome editing is a type of genetic engineering in which DNA is inserted, replaced, or removed from a genome using artificially engineered nucleases, or "molecular scissors." The nucleases create specific double-stranded breaks (DSBs) at desired locations in the genome, and harness the cell's endogenous mechanisms to repair the induced break by natural processes of homologous recombination (HR) and nonhomologous end-joining (NHEJ). There are currently four families of engineered nucleases: meganucleases, zinc finger nucleases (ZFNs), transcription activator-like effector nucleases (TALENs), and the Cas9-guideRNA system (adapted from the CRISPR prokarotic immune system). In contrast to artificial genome editing natural genome editing occurs through viral and sub-viral agents competent in identification of genetic syntax structures for insertion/deletion processes with the result of conserved selection processes.

Applications

Genetic engineering has applications in medicine, research, industry and agriculture and can be used on a wide range of plants, animals and micro organisms.

Medicine

In medicine, genetic engineering has been used in manufacturing drugs, to create model animals and do laboratory research, and in gene therapy.

Manufacturing

Genetic engineering is used to mass-produce insulin, human growth hormones, follistim (for treating infertility), human albumin, monoclonal antibodies, antihemophilic factors, vaccines and many other drugs. Mouse hybridomas, cells fused together to create monoclonal antibodies, have been humanised through genetic engineering to create human monoclonal antibodies. Genetically engineered viruses are being developed that can still confer immunity, but lack the infectious sequences.

Research

Genetic engineering is used to create animal models of human diseases. Genetically modified mice are the most common genetically engineered animal model. They have been used to study and model cancer (the oncomouse), obesity, heart disease, diabetes, arthritis, substance abuse, anxiety, aging and Parkinson disease. Potential cures can be tested against these mouse models. Also genetically modified pigs have been bred with the aim of increasing the success of pig to human organ transplantation.

Gene Therapy

Gene therapy is the genetic engineering of humans, generally by replacing defective genes with effective ones. This can occur in somatic tissue or germline tissue.

Somatic gene therapy has been studied in clinical research in several diseases, including X-linked SCID, chronic lymphocytic leukemia (CLL), and Parkinson's disease. In 2012, Glybera became the first gene therapy treatment to be approved for clinical use in either Europe or the United States after its endorsement by the European Commission.

With regard to germline gene therapy, the scientific community has been opposed to attempts to alter genes in humans in inheritable ways using biotechnology since the technology was first introduced, and the caution has continued as the technology has progressed. With the advent of new techniques like CRISPR, in March 2015 scientists urged a worldwide ban on clinical use of gene editing technologies to edit the human genome in a way that can be inherited. In April 2015, Chinese researchers sparked controversy when they reported results of basic research experiments in which they edited the DNA of non-viable human embryos using CRISPR. In December 2015, scientists of major world academies called for a moratorium on inheritable human genome edits, including those related to CRISPR-Cas9 technologies.

There are also ethical concerns should the technology be used not just for treatment, but for enhancement, modification or alteration of a human beings' appearance, adaptability, intelligence, character or behavior. The distinction between cure and enhancement can also be difficult to establish. Transhumanists consider the enhancement of humans desirable.

Research

Knockout mice

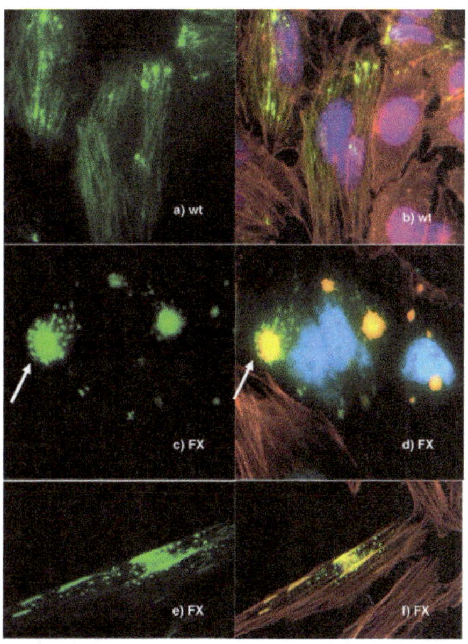

Human cells in which some proteins are fused with green fluorescent protein to allow them to be visualised

Genetic engineering is an important tool for natural scientists. Genes and other genetic information from a wide range of organisms are transformed into bacteria for storage and modification, creating genetically modified bacteria in the process. Bacteria are cheap, easy to grow, clonal, multiply quickly, relatively easy to transform and can be stored at -80 °C almost indefinitely. Once a gene is isolated it can be stored inside the bacteria providing an unlimited supply for research.

Organisms are genetically engineered to discover the functions of certain genes. This could be the effect on the phenotype of the organism, where the gene is expressed or what other genes it interacts with. These experiments generally involve loss of function, gain of function, tracking and expression.

- Loss of function experiments, such as in a gene knockout experiment, in which an organism is engineered to lack the activity of one or more genes. A knockout experiment involves the creation and manipulation of a DNA construct *in vitro*, which, in a simple knockout, consists of a copy of the desired gene, which has been altered such that it is non-functional. Embryonic stem cells incorporate the altered gene, which replaces the already present functional copy. These stem cells are injected into blastocysts, which are implanted into surrogate mothers. This allows the experimenter to analyze the defects caused by this mutation and thereby determine the role of particular genes. It is used especially frequently in developmental biology. Another method, useful in organisms such as Drosophila (fruit fly), is to induce mutations in a large population and then screen the progeny for the desired mutation. A similar process can be used in both plants and prokaryotes. Loss of function tells whether or not a protein is required for a function, but it does not always mean it's sufficient, especially if a function requires multiple proteins and lose the said function if one protein is missing.

- Gain of function experiments, the logical counterpart of knockouts. These are sometimes performed in conjunction with knockout experiments to more finely establish the function of the desired gene. The process is much the same as that in knockout engineering, except that the construct is designed to increase the function of the gene, usually by providing extra copies of the gene or inducing synthesis of the protein more frequently. Gain of function is used to tell whether or not a protein is sufficient for a function, but it does not always mean it's required. Especially when dealing with genetic/functional redundancy.

- Tracking experiments, which seek to gain information about the localization and interaction of the desired protein. One way to do this is to replace the wild-type gene with a 'fusion' gene, which is a juxtaposition of the wild-type gene with a reporting element such as green fluorescent protein (GFP) that will allow easy visualization of the products of the genetic modification. While this is a useful technique, the manipulation can destroy the function of the gene, creating secondary effects and possibly calling into question the results of the experiment. More sophisticated techniques are now in development that can track protein products without mitigating their function, such as the addition of small sequences that will serve as binding motifs to monoclonal antibodies.

- Expression studies aim to discover where and when specific proteins are produced. In these experiments, the DNA sequence before the DNA that codes for a protein, known as a gene's promoter, is reintroduced into an organism with the protein coding region replaced by a reporter gene such as GFP or an enzyme that catalyzes the production of a dye. Thus the time and place where a particular protein is produced can be observed. Expression studies can be taken a step further by altering the promoter to find which pieces are crucial for the proper expression of the gene and are actually bound by transcription factor proteins; this process is known as promoter bashing.

Industrial

Using genetic engineering techniques one can transform microorganisms such as bacteria or yeast, or transform cells from multicellular organisms such as insects or mammals, with a gene coding for a useful protein, such as an enzyme, so that the transformed organism will overexpress the desired protein. One can manufacture mass quantities of the protein by growing the transformed organism in bioreactor equipment using techniques of industrial fermentation, and then purifying the protein. Some genes do not work well in bacteria, so yeast, insect cells, or mammalians cells, each a eukaryote, can also be used. These techniques are used to produce medicines such as insulin, human growth hormone, and vaccines, supplements such as tryptophan, aid in the production of food (chymosin in cheese making) and fuels. Other applications involving genetically engineered bacteria being investigated involve making the bacteria perform tasks outside their natural cycle, such as making biofuels, cleaning up oil spills, carbon and other toxic waste and detecting arsenic in drinking water. Certain genetically modified microbes can also be used in biomining and bioremediation, due to their ability to extract heavy metals from their environment and incorporate them into compounds that are more easily recoverable.

Experimental, Lab Scale Industrial Applications

In materials science, a genetically modified virus has been used in an academic lab as a scaffold for assembling a more environmentally friendly lithium-ion battery.

Bacteria have been engineered to function as sensors by expressing a fluorescent protein under certain environmental conditions.

Agriculture

One of the best-known and controversial applications of genetic engineering is the creation and use of genetically modified crops or genetically modified organisms, such as genetically modified fish, which are used to produce genetically modified food and materials with diverse uses. There are four main goals in generating genetically modified crops.

One goal, and the first to be realized commercially, is to provide protection from environmental threats, such as cold (in the case of Ice-minus bacteria), or pathogens, such as insects or viruses, and/or resistance to herbicides. There are also fungal and virus resistant crops developed or in development. They have been developed to make the insect and weed management of crops easier and can indirectly increase crop yield.

Bt-toxins present in peanut leaves (bottom image) protect it from extensive damage caused by European corn borer larvae (top image).

Another goal in generating GMOs is to modify the quality of produce by, for instance, increasing the nutritional value or providing more industrially useful qualities or quantities. The Amflora potato, for example, produces a more industrially useful blend of starches. Cows have been engineered to produce more protein in their milk to facilitate cheese production. Soybeans and canola have been genetically modified to produce more healthy oils.

Another goal consists of driving the GMO to produce materials that it does not normally make. One example is "pharming", which uses crops as bioreactors to produce vaccines, drug intermediates, or drug themselves; the useful product is purified from the harvest and then used in the standard pharmaceutical production process. Cows and goats have been engineered to express drugs and other proteins in their milk, and in 2009 the FDA approved a drug produced in goat milk.

Another goal in generating GMOs, is to directly improve yield by accelerating growth, or making the organism more hardy (for plants, by improving salt, cold or drought tolerance). Salmon have been genetically modified with growth hormones to increase their size.

The genetic engineering of agricultural crops can increase the growth rates and resistance to different diseases caused by pathogens and parasites. This is beneficial as it can greatly increase the production of food sources with the usage of fewer resources that would be required to host the world's growing populations. These modified crops would also reduce the usage of chemicals, such as fertilizers and pesticides, and therefore decrease the severity and frequency of the damages produced by these chemical pollution.

Ethical and safety concerns have been raised around the use of genetically modified food. A major safety concern relates to the human health implications of eating genetically modified food, in particular whether toxic or allergic reactions could occur. Gene flow into related non-transgenic crops, off target effects on beneficial organisms and the impact on biodiversity are important environmental issues. Ethical concerns involve religious issues, corporate control of the food supply, intellectual property rights and the level of labeling needed on genetically modified products.

Conservation

Genetic engineering has potential applications in conservation and natural areas management. For example, gene transfer through viral vectors has been proposed as a means of controlling invasive species as well as vaccinating threatened fauna from disease. Transgenic trees have also been suggested as a way to confer resistance to pathogens in wild populations. With the increasing risks of maladaptation in organisms as a result of climate change and other perturbations, facilitated adaptation through gene tweaking could be one solution to reducing extinction risks. Applications of genetic engineering in conservation are thus far mostly theoretical and have yet to be put into practice. Further experimentation will be necessary to gauge the benefits and costs of such practices.

BioArt and Entertainment

Genetic engineering is also being used to create BioArt. Some bacteria have been genetically engineered to create black and white photographs.

Genetic engineering has also been used to create novelty items such as lavender-colored carnations, blue roses, and glowing fish.

Regulation

The regulation of genetic engineering concerns the approaches taken by governments to assess and manage the risks associated with the development and release of genetically modified crops. There are differences in the regulation of GM crops between countries, with some of the most marked differences occurring between the USA and Europe. Regulation varies in a given country depending on the intended use of the products of the genetic engineering. For example, a crop not intended for food use is generally not reviewed by authorities responsible for food safety. Starting in the late 1980s, guidance on assessing the safety of genetically engineered plants and food emerged from organizations including the FAO and WHO.

Controversy

Critics have objected to use of genetic engineering per se on several grounds, including ethical concerns, ecological concerns, and economic concerns raised by the fact GM techniques and GM organisms are subject to intellectual property law. GMOs also are involved in controversies over GM food with respect to whether food produced from GM crops is safe, whether it should be labeled, and whether GM crops are needed to address the world's food needs. See the genetically modified food controversies article for discussion of issues about GM crops and GM food. These controversies have led to litigation, international trade disputes, and protests, and to restrictive regulation of commercial products in some countries.

Genetic Recombination

nation, initiated by a double-strand break or gap, followed by pairing with an homologous chromosome and strand invasion to initiate the recombinational repair process. Repair of the gap can lead to crossover (CO) or non-crossover (NCO) of the flanking regions. CO recombination is thought to occur by the Double Holliday Junction (DHJ) model, illustrated on the right, above. NCO recombinants are thought to occur primarily by the Synthesis Dependent Strand Annealing (SDSA) model, illustrated on the left, above. Most recombination events appear to be the SDSA type.

Genetic recombination is the production of offspring with combinations of traits that differ from those found in either parent. In eukaryotes, genetic recombination during meiosis can lead to a novel set of genetic information that can be passed on from the parents to the offspring. Most recombination is naturally occurring.

During meiosis in eukaryotes, genetic recombination involves the pairing of homologous chromosomes. This may be followed by information transfer between the chromosomes. The information transfer may occur without physical exchange (a section of genetic material is copied from one chromosome to another, without the donating chromosome being changed) or by the breaking and rejoining of DNA strands, which forms new molecules of DNA DHJ pathway in Figure.

Recombination may also occur during mitosis in eukaryotes where it ordinarily involves the two sister chromosomes formed after chromosomal replication. In this case, new combinations of alleles are not produced since the sister chromosomes are usually identical. In meiosis and mitosis, recombination occurs between similar molecules of DNA (homologs). In meiosis, non-sister homologous chromosomes pair with each other so that recombination characteristically occurs between non-sister homologues. In both meiotic and mitotic cells, recombination between homologous chromosomes is a common mechanism used in DNA repair.

Genetic recombination and recombinational DNA repair also occurs in bacteria and archaea, which use asexual reproduction.

Recombination can be artificially induced in laboratory (in vitro) settings, producing recombinant DNA for purposes including vaccine development.

V(D)J recombination in organisms with an adaptive immune system is a type of site-specific genetic recombination that helps immune cells rapidly diversify to recognize and adapt to new pathogens.

Synapsis

During meiosis, synapsis (the pairing of homologous chromosomes) ordinarily precedes genetic recombination.

Mechanism

Genetic recombination is catalyzed by many different enzymes. Recombinases are key enzymes that catalyse the strand transfer step during recombination. RecA, the chief recombinase found in *Escherichia coli*, is responsible for the repair of DNA double strand breaks (DSBs). In yeast and other eukaryotic organisms there are two recombinases required for repairing DSBs. The RAD51 protein is required for mitotic and meiotic recombination, whereas the DNA repair protein, DMC1, is specific to meiotic recombination. In the archaea, the ortholog of the bacterial RecA protein is RadA.

Chromosomal Crossover

In eukaryotes, recombination during meiosis is facilitated by chromosomal crossover. The crossover process leads to offspring having different combinations of genes from those of their parents, and can occasionally produce new chimeric alleles. The shuffling of genes brought about by genetic recombination produces increased genetic variation. It also allows sexually reproducing organisms to avoid Muller's ratchet, in which the genomes of an asexual population accumulate genetic deletions in an irreversible manner.

Chromosomal crossover involves recombination between the paired chromosomes inherited from each of one's parents, generally occurring during meiosis. During prophase I (pachytene stage) the four available chromatids are in tight formation with one another. While in this formation, homologous sites on two chromatids can closely pair with one another, and may exchange genetic information.

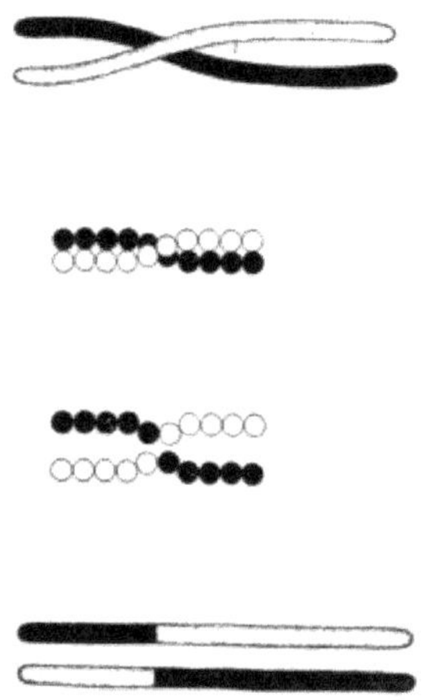

FIG. 64. Scheme to illustrate a method of crossing over of the chromosomes.

Thomas Hunt Morgan's illustration of crossing over (1916)

Because recombination can occur with small probability at any location along chromosome, the frequency of recombination between two locations depends on the distance separating them. Therefore, for genes sufficiently distant on the same chromosome the amount of crossover is high enough to destroy the correlation between alleles.

Tracking the movement of genes resulting from crossovers has proven quite useful to geneticists. Because two genes that are close together are less likely to become separated than genes that are farther apart, geneticists can deduce roughly how far apart two genes are on a chromosome if they know the frequency of the crossovers. Geneticists can also use this method to infer the presence of certain genes. Genes that typically stay together during recombination are said to be linked. One gene in a linked pair can sometimes be used as a marker to deduce the presence of another gene. This is typically used in order to detect the presence of a disease-causing gene.

Gene Conversion

In gene conversion, a section of genetic material is copied from one chromosome to another, without the donating chromosome being changed. Gene conversion occurs at high frequency at the actual site of the recombination event during meiosis. It is a process by which a DNA sequence is copied from one DNA helix (which remains unchanged) to another DNA helix, whose sequence is altered. Gene conversion has often been studied in fungal crosses where the 4 products of individual meioses can be conveniently observed. Gene conversion events can be distinguished as deviations in an individual meiosis from the normal 2:2 segregation pattern (e.g. a 3:1 pattern).

Nonhomologous Recombination

Recombination can occur between DNA sequences that contain no sequence homology. This can cause chromosomal translocations, sometimes leading to cancer.

In B Cells

B cells of the immune system perform genetic recombination, called immunoglobulin class switching. It is a biological mechanism that changes an antibody from one class to another, for example, from an isotype called IgM to an isotype called IgG.

Genetic Engineering

In genetic engineering, recombination can also refer to artificial and deliberate recombination of disparate pieces of DNA, often from different organisms, creating what is called recombinant DNA. A prime example of such a use of genetic recombination is gene targeting, which can be used to add, delete or otherwise change an organism's genes. This technique is important to biomedical researchers as it allows them to study the effects of specific genes. Techniques based on genetic recombination are also applied in protein engineering to develop new proteins of biological interest.

Recombinational Repair

During both mitosis and meiosis, DNA damages caused by a variety of exogenous agents (e.g. UV light, X-rays, chemical cross-linking agents) can be repaired by homologous recombinational repair (HRR). These findings suggest that DNA damages arising from natural processes, such as exposure to reactive oxygen species that are byproducts of normal metabolism, are also repaired by HRR. In humans and rodents, deficiencies in the gene products necessary for HRR during meiosis cause infertility. In humans, deficiencies in gene products necessary for HRR, such as BRCA1 and BRCA2, increase the risk of cancer DNA repair-deficiency disorder.

In bacteria, transformation is a process of gene transfer that ordinarily occurs between individual cells of the same bacterial species. Transformation involves integration of donor DNA into the recipient chromosome by recombination. This process appears to be an adaptation for repairing DNA damages in the recipient chromosome by HRR. Transformation may provide a benefit to pathogenic bacteria by allowing repair of DNA damage, particularly damages that occur in the inflammatory, oxidizing environment associated with infection of a host.

When two or more viruses, each containing lethal genomic damages, infect the same host cell, the virus genomes can often pair with each other and undergo HRR to produce viable progeny. This process, referred to as multiplicity reactivation, has been studied in bacteriophages T4 and lambda, as well as in several pathogenic viruses. In the case of pathogenic viruses, multiplicity reactivation may be an adaptive benefit to the virus since it allows the repair of DNA damages caused by exposure to the oxidizing environment produced during host infection.

Meiotic Recombination

Molecular models of meiotic recombination have evolved over the years as relevant evidence accumulated. A major incentive for developing a fundamental understanding of the mechanism of meiotic recombination is that such understanding is crucial for solving the problem of the adaptive function of sex, a major unresolved issue in biology. A recent model that reflects current understanding was presented by Anderson and Sekelsky, and is outlined in the first figure in this article. The figure shows that two of the four chromatids present early in meiosis (prophase I) are paired with each other and able to interact. Recombination, in this version of the model, is initiated by a double-strand break (or gap) shown in the DNA molecule (chromatid) at the top of the first figure in this article. However, other types of DNA damage may also initiate recombination. For instance, an inter-strand cross-link (caused by exposure to a cross-linking agent such as mitomycin C) can be repaired by HRR.

As indicated in the first figure, above, two types of recombinant product are produced. Indicated on the right side is a "crossover" (CO) type, where the flanking regions of the chromosomes are exchanged, and on the left side, a "non-crossover" (NCO) type where the flanking regions are not exchanged. The CO type of recombination involves the intermediate formation of two "Holliday junctions" indicated in the lower right of the figure by two X shaped structures in each of which there is an exchange of single strands between the two participating chromatids. This pathway is labeled in the figure as the DHJ (double-Holliday junction) pathway.

The NCO recombinants (illustrated on the left in the figure) are produced by a process referred to as "synthesis dependent strand annealing" (SDSA). Recombination events of the NCO/SDSA type appear to be more common than the CO/DHJ type. The NCO/SDSA pathway contributes little to genetic variation since the arms of the chromosomes flanking the recombination event remain in the parental configuration. Thus, explanations for the adaptive function of meiosis that focus exclusively on crossing-over are inadequate to explain the majority of recombination events.

Achiasmy and Heterochiasmy

Achiasmy is the phenomenon where autosomal recombination is completely absent in one sex of a species. Achiasmatic chromosomal segregation is well documented in male *Drosophila melanogaster*. Heterochiasmy is the term used to describe recombination rates which differ between the sexes of a species. This sexual dimorphic pattern in recombination rate has been observed in many species. In mammals, females most often have higher rates of recombination. The "Haldane-Huxley rule" states that achiasmy usually occurs in the heterogametic sex.

Family-based QTL Mapping

Quantitative trait loci mapping or QTL mapping is the process of identifying genomic regions that potentially contain genes responsible for important economic, health or environmental characters. Mapping QTLs is an important activity that plant breeders and geneticists routinely use to associate potential causal genes with phenotypes of interest. Family-based QTL mapping is a variant of QTL mapping where multiple-families are used.

Pedigree in Humans and Wheat

Pedigree information include information about ancestry. Keeping pedigree records is a centuries-old tradition. Pedigrees can also be verified using gene-marker data.

A complex five generation plant pedigree drawn with Pedimap

Family-based QTL Mapping in Plants

The method has been discussed in the context of plant breeding populations. Pedigree records are kept by plants breeders and pedigree-based selection is popular in several plant species. Plant pedigrees are different from that of humans, particularly as plant are hermaphroditic - an individual can be male or female and mating can be performed in random combinations, with inbreeding loops. Also plant pedigrees may contain of "selfs", i.e. cloned offspring resulting from self-pollination of a plant.

Pedigree Denotation

SIMPLE CROSS SYMBOL	Example
/ first order cross	SON 64/KLRE
//, second order cross	IR 64/KLRE // CIANo
/3/, third order cross	TOBS /3/ SON 64/KLRE // CIANo
/4/, fourth order cross	TOBS /3/ SON 64/KLRE // CIANo /4/
/n/, nth order cross	
BACK CROSS SYMBOL	
*n	n number of times the back cross parent used left side simple cross symbol, back cross parent is the female, right side – male,

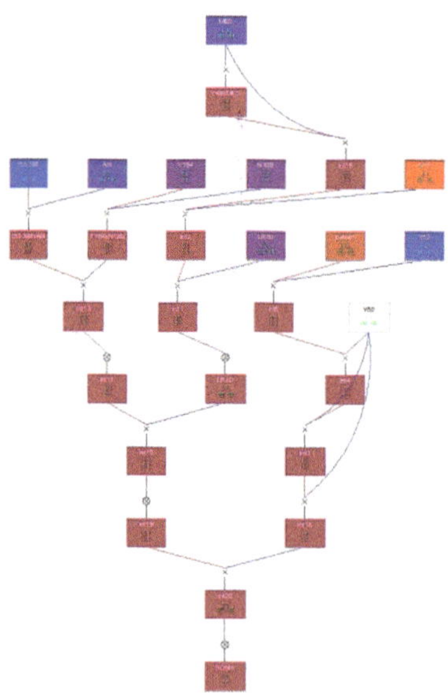

Example pedigree of Sonalika (SONALIKA = = I53.388/AN//YT54/N10B/3/Lerma Rojo/4/ B4946.A.4.18.2.IY/Y53//3*Y50 drawn using Pedimap

The idea of family-based QTL mapping comes from inheritance of marker alleles and its association with trait of interest has demonstrated how to use family-based association in plant breeding families.

Limitation of Conventional Methods

Traditional mapping populations include single family consisting of crossing between two parents or three parents often distantly related. There are some important limitations associated with traditional mapping methods. Some of which include limited polymorphism rates, and no indication of marker effectiveness in multiple genetic backgrounds. Often, by the time a QTL mapping population is developed and mapped, breeders have introgressed the new QTL using traditional breeding and selection methods. This can reduce the usefulness of MAS (marker-assisted selection) within breeding programs at the time when MAS could be most useful (i.e., shortly after new QTL are identified). Family-based QTL mapping removes this limitation by using existing plant breeding families.

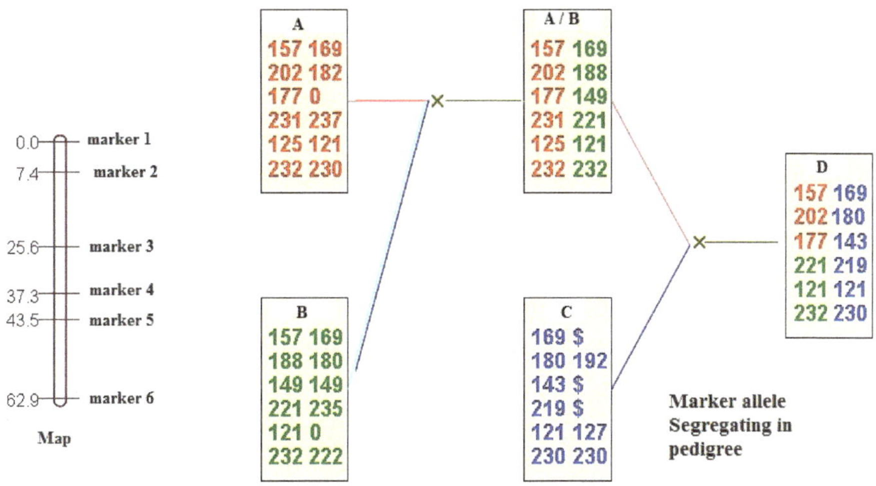

Common Study Population Mapping

Broadly, there are 3 classes of study designs: study designs in which large sets of relatives from extended or nuclear families are sampled, study designs in which pairs of relatives are sampled (e.g., sibling pairs) or study designs in which unrelated individuals are sampled.

Unrelated Individuals

Natural collection of individuals (considered unrelated) with unknown pedigree constitutes mapping populations. The population based association mapping technique are based on this type of populations. In plant context such population are hard to find as most of individuals are someway related. Other disadvantage of such method is that even if we can find such a population, it is difficult to find high allele frequency for allele of interest (usually mutant)in such situation. For purpose of create balance in allele frequency, usually case-control studies.

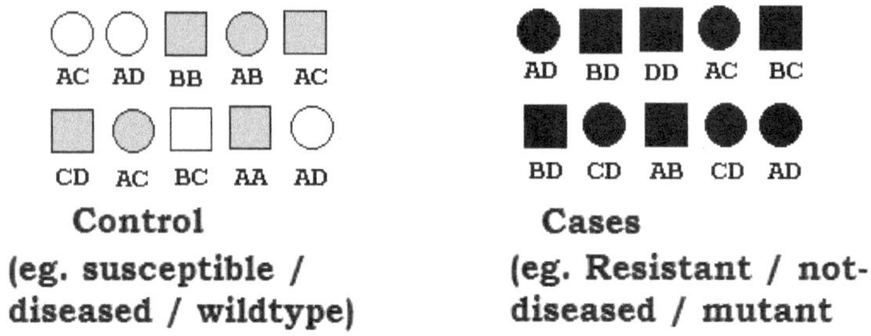

Sibpairs

Such design include a pair of sibs from multiple independent families. The members in each sibpairs are not randomly chosen - often both siblings are chosen from one tail (upper or lower) of the distribution of the QT (concordant siblings) or one sibling is chosen from the upper tail and the other sibling is chosen from the lower tail (discordant siblings). Another sampling design could include a pair of siblings, one chosen from the upper or lower tail of the distribution and the other chosen randomly from among the remaining siblings.

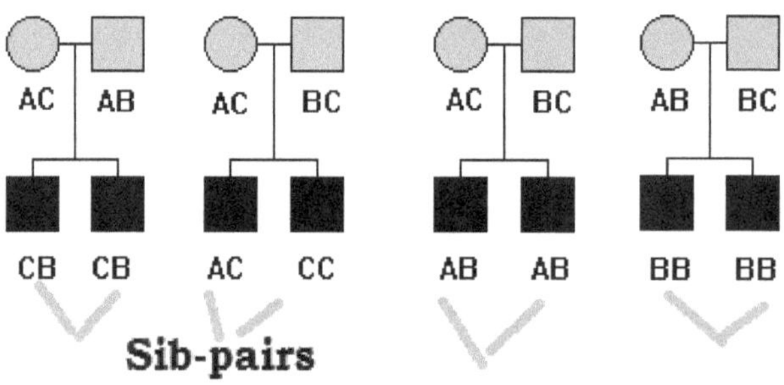

Trios

Trios include parents and one offspring (most affected). Trios are more commonly used in association studies. The concept of association mapping that each trio are unrelated, however trios are related in themselves.

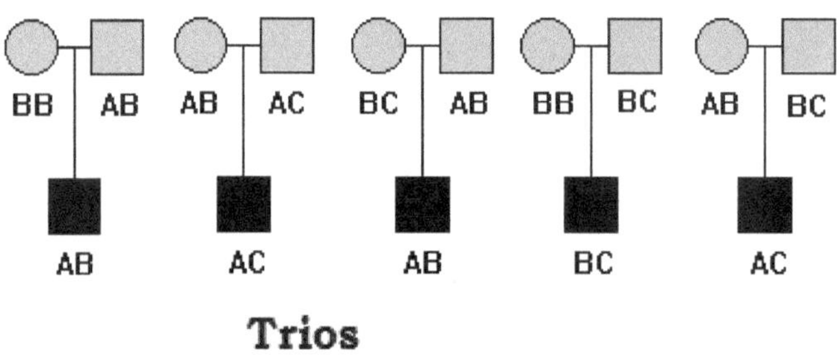

Nuclear Family

Nuclear family consists of two generation simple family pedigree.

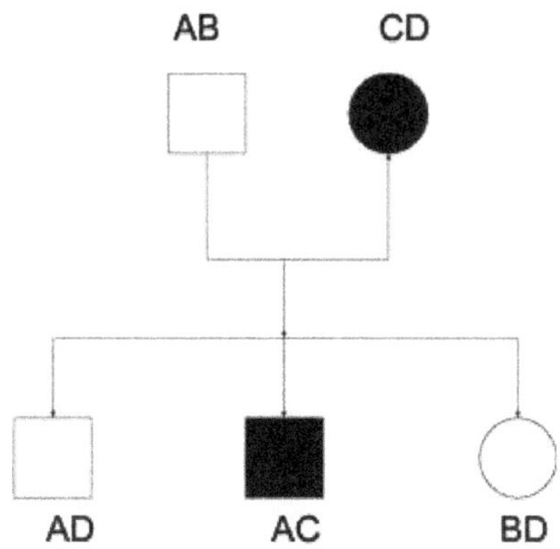

Extended Pedigrees

In extended pedigree include multiple generation pedigree. It can be as deep or wide as the pedigree information is available. Extended pedigree are attractive for linkage-based analysis.

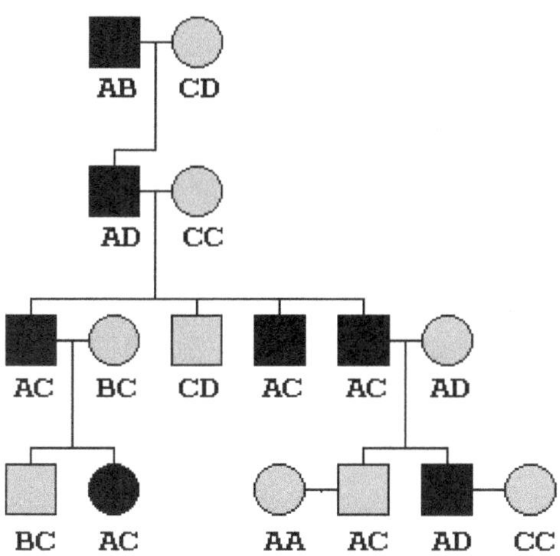

Linkage vs Association Analysis

Linkage and association analysis are primary tools for gene discovery, localization and functional analysis. While conceptual underpinning of these approaches have been long known, advances in recent decades in molecular genetics, development in efficient algorithms, and computing power have enabled the large scale application of these methods. While linkage studies seek to identify loci cosegregate with the trait within families, association studies seek to identify particular variants that are associated with the phenotype at the population level. These are complementary methods that, together, provide means to probe the genome and describe etiology of complex traits. In linkage studies, we seek to identify the loci that cosegregate with a specific genomic region, tagged by polymorphic markers, within families. In contrast, in association studies, we seek a correlation between a specific genetic variation and trait variation in sample of individuals, implicating a causal role of the variant.

Family-based Linkage Analysis

Genetic linkage is the phenomenon where by alleles at different loci cosegregate in families. The strength of cosegregation is measured by the recombination fraction θ, the probability of an odd number of recombination. More complex pedigree provide higher power. Identity by descent (IBD) matrix estimation is a central component in mapping of Quantitative Trait Loci (QTL) using variance component models. Alleles have identity by type (IBT) when they have the same phenotypic effect. Alleles that are identical by type fall into two groups; those that are identical by descent (IBD) because they arose from the same allele in an earlier generation; and those that are non-identical by descent (NIBD) or identical by state (IBS) because they arose from separate mutations. Parent-offspring pairs share 50% of their genes IBD, and monozygotic twins share 100% IBD. It should be noted that what is relevant in linkage analysis is the inheritance (or coinheritance) of alleles at adjacent loci; therefore; it is critical importance to determine whether the alleles are identical by descent (i.e. copies from same parental alleles) or only identical by state (i.e. appearing same, but derived from two different copies of alleles). Therefore there three categories of family-based linkage analysis – strongly modeled (the traditional lod score model), weakly model based (variance components methods), or model free. Variance component methods may be viewed as hybrids.

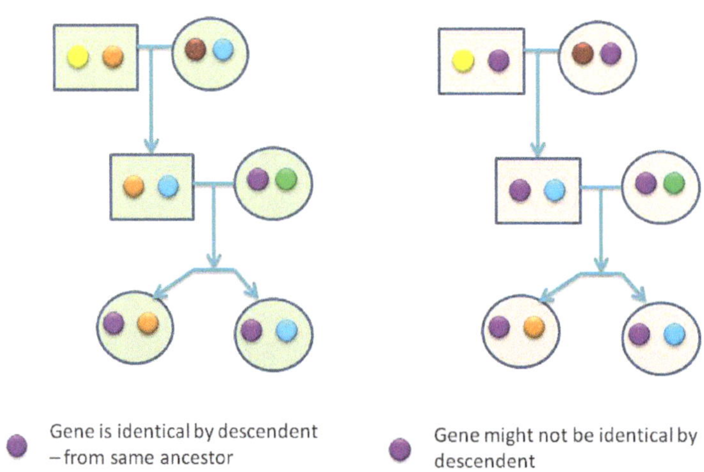

IBD concept

Family-based Association Analysis

Linkage disequilibrium (LD) and association mapping is receiving considerable attention in the plant genetics community for its potential to use existing genetic resources collections to fine map quantitative trait loci (QTL), validate candidate genes, and identify alleles of interest (Yu and Buckler, 2006). The three elements of particular importance for conducting association mapping or interpreting the results include:

1. the analysis of population structure into subgroups,
2. its use to control for spurious associations and consequences in the specific case of differential selection among subgroups, and
3. the analysis of the local structure of LD into haplotypes and its consequences on the resolution and the application of LD mapping (Flint-Garcia et al. 2003).

In contrast to population-based association, family-based association tests are becoming more popular.

The family-based, Tran-disequilibirum test (TDT) has gained wide popularity in recent years, this method also focuses on alleles transmitted to affect offispring, but it is formulated to take account of both the linkage and the disequilibrium that underlie the association. The test requires genotype information on trio individuals, namely affected child and both biological parents; and at least one parent must be heterozygous for the test to be informative. The proposed test statistic is actually McNemar's chi-square statistic and tests the null hypothesis that the putative disease associated allele is transmitted 50% of the time from the heterogygous parents against the alternative hypothesis that the trait positive allele -associated allele will be transmitted more often. The TDT is not affected by population stratification and admixture. The concept of family-based test of association has been extended to quantitative traits.

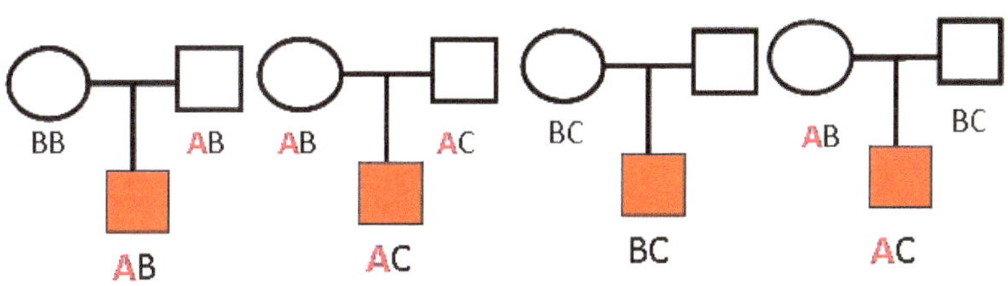

The A allele is transmitted to resistant or susceptible offspring three times out of four

Quantitative Transmission Disequilibrium Test (QTDT)

The TDT has been extended in context of quantitative traits and nuclear or extended pedigree families. The generalized test allows to use any family type of families in testing. QTDT has also be been extended to haplotype-based association mapping. Haplotypes refer to combinations of marker alleles which are located closely together on the same chromosome and which tend to be inherited together. With availability of high density SNP makers, haplotypes play an important role in association studies. First – haplotypes are critical to understanding the LD pattern across the genome, which is essential for association studies. Actually there is no better way to understand LD pattern than to know the haplotypes themselves. Haplotypes tell us how alleles are organized along the chromosome and reflect the pattern of inheritance over evaluations. Second, methods based on haplotypes can be more powerful than those based on single markers in association studies of mapping complex trait genes.

Drawing Family Pedigrees

There are several pedigree drawing software available for human genetics context such as COPE (COllaborative Pedigree drawing Environment), CYRILLIC, FTM (Family Tree Maker), FTREE, KINDRED, PED (PEdigree Drawing software),PED-HUNTER, PEDIGRAPH, PEDIGREE/DRAW, PEDIGREE-VISUALIZER, PED-PLOT,PEDRAW/WPEDRAW (Pedigree Drawing/ Window Pedigree Drawing (MS-Window and X-Window version of PEDRAW)), PROGENY (Progeny Software, LLC) etc. However the pedigree drawing in plants requires some additional features such as inbreeding, selfing, mutation, polyploidy etc. which is supported in Pedimap. The pedimap can be used for pedigree visualization along with phenotypic, genotypic and ibd probabilities data in every type of plant pedigrees in both diploids and tetraploids.

Breeding for Drought Stress Tolerance

Breeding for drought stress tolerance is the process of breeding plants with the goal of reducing the impact of drought on plant growth.

Drought Stress in Crop Plants

In nature or crop fields, water is often the most limiting factor for plant growth. If plants do not receive adequate rainfall or irrigation, the resulting drought stress can reduce growth more than all other environmental stresses combined.

Drought can be defined as the absence of rainfall or irrigation for a period of time sufficient to deplete soil moisture and injure plants. Drought stress results when water loss from the plant exceeds the ability of the plant's roots to absorb water and when the plant's water content is reduced enough to interfere with normal plant processes.

Drought Stress is a Global Phenomenon

About 15 million km2 of the land surface is covered by cropland (), and about 16% of this area is equipped for irrigation (Siebert *et al.* 2005). Thus in many parts of the world including the United States, plants may frequently encounter drought stress. Rainfall is very seasonal and periodic drought occurs regularly. The effect of drought is more prominent in sandy soils with low water holding capacity. On such soils some plants may experience drought stress after only a few days without water.

During the last century, the rate of increase in `blue' water withdrawal (from rivers, lakes, and aquifers) for irrigation and other purposes was higher than the growth rate of the world population (Shiklomanov 1998). Country-wise maps of irrigated areas are available.

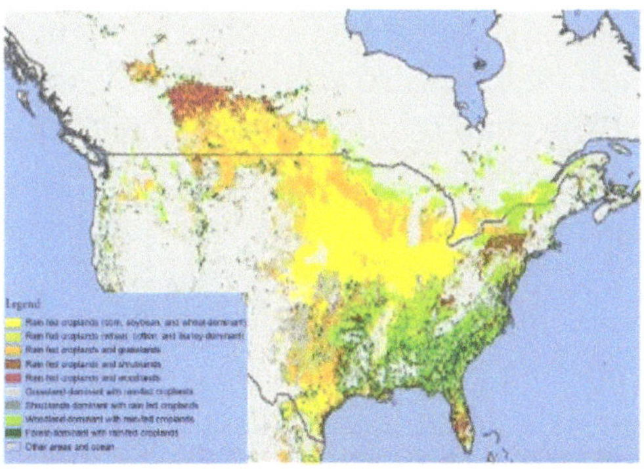

Rain fed areas of USA, for details - Biradar et al.(2009) International J Applied Earth Observation and Geoinformation 11: 114-129

Drought Stress and Future Challenges to Crop Production

Moisture deficits is a significant challenge to the future crop production. Severe drought in parts of the U.S., Australia, and Africa in recent years drastically reduced crop yields and disrupted regional economies. Even in average years, however, many agricultural regions, including the U.S. Great Plains, suffer from chronic moisture deficits. Cereal crops typically attain only about 25% of their potential yield due to the effects of environmental stress, with moisture stress the most important cause. Two major trends will likely increase the frequency and severity of crop moisture deficits:

(1) Global Climate Change.

Higher temperatures are likely to increase crop water use due to increased transpiration. A warmer atmosphere will also speed up melting of mountain snowpack, resulting in less water available for irrigation. More extreme weather patterns will increase the frequency of drought in some regions.

(2) Competing uses for Limited Water Supplies.

Increased demand from municipal and industrial users will further reduce the amount of water available for irrigated crops.

Although changes in tillage and irrigation practices can improve production by conserving water, enhancing the genetic tolerance of crops to drought stress is considered an essential strategy for addressing moisture deficits.

Drought Stress Affects Plant Physiology

A plant responds to a lack of water by halting growth and reducing photosynthesis and other plant processes in order to reduce water use. As water loss progresses, leaves of some species may appear to change color—usually to blue-green. Foliage begins to wilt and, if the plant is not irrigated, leaves will fall off and the plant will eventually die. Drought lowers the water potential of a plant's root and upon extended exposure, abscisic acid is accumulated and eventually stomatal closure occurs. This reduces a plant's leaf relative water content. The time required for drought stress to occur depends on the water-holding capacity of the soil, environmental conditions, stage of plant growth, and plant species. Plants growing in sandy soils with low water-holding capacity are more susceptible to drought stress than plants growing in clay soils. A limited root system will accelerate the rate at which drought stress develops. A root system may be limited by the presence of competing root systems, by site conditions such as compacted soils or high water tables, or by container size (if growing in a container). A plant with a large mass of leaves in relation to the root system is prone to drought stress because the leaves may lose water faster than the roots can supply it. Newly installed plants and poorly established plants may be especially susceptible to drought stress because of the limited root system or the large mass of stems and leaves in comparison to roots.

Drought Stress Interaction with Other Stress Factors

Aside from the moisture content of the soil, environmental conditions of high light intensity, high temperature, low relative humidity and high wind speed will significantly increase plant water loss. The prior environment of a plant also can influence the development of drought stress. A plant that has been drought stressed previously and has recovered may become more drought resistant. Also, a plant that was well-watered prior to drought will usually survive drought better than a continuously drought-stressed plant.

Mechanism of Drought Stress Resistance

Degree of Resistance to drought depend upon crops. Generally three strategies can help a crop to mitigate the effect of drought stress:

(a) drought escape

(b) drought avoidance

(c) drought tolerance

A proper timing of lifecycle, resulting in the completion of the most sensitive developmental stages while water is abundant, is considered to be a drought escape strategy. Avoiding water-deficit stress with a root system capable of extracting water from deep soil layers, or by reducing evapotranspiration without affecting yields, is considered as drought avoidance. Mechanisms such as osmotic adjustment (OA) whereby a plant maintains cell turgor pressure under reduced soil water potential are categorized as drought tolerance mechanisms. Drought avoidance mechanisms can be expressed even in the absence of stress and are then considered constitutive. Drought tolerance mechanisms are the result of a response triggered by drought stress itself and are therefore considered adaptative. When the stress is terminal and predictable, drought escape through the use of shorter duration varieties is often the preferable method of improving yield potential. Drought avoidance and tolerance mechanisms are required in situations where the timing of drought is mostly unpredictable.

Drought tolerance mechanism is genetically controlled and genes or QTL responsible for drought tolerance have been discovered in several crops which opens avenue for molecular breeding for drought tolerance.

Drought Tolerance Traits

Tolerance to drought is a quantitative trait, with a complex phenotype, often confounded by plant phenology. Breeding for drought tolerance is further complicated since several types of abiotic stress, such as high temperatures, high irradiance, and nutrient toxicities or deficiencies can challenge crop plants simultaneously.

(a) Osmotic adjustment When a plant is exposed to water deficit, it may accumulate a variety of osmotically active compounds such as amino acids and sugars, resulting in a lowering of the osmotic potential. Examples of amino acids that may be up-regulated are proline and glycine betaine. This is termed osmotic adjustment and enables the plant to take up water, maintain turgor and survive longer.

(b) Cell membrane stability The ability to survive dehydration is influenced by a cell's ability to survive at reduced water content. This can be considered complementary to OA because both traits will help maintain leaf growth (or prevent leaf death) during drought. Crop varieties differ in dehydration tolerance and an important factor for such differences is the capacity of the cell membrane to prevent electrolyte leakage at decreasing water content, or "cell membrane stability (CMS)". The maintenance of membrane function is assumed to mean that cell activity is also maintained. Measurements of CMS have been used in different crops and are known to be correlated with yields under high temperature and possibly under drought stress.

(c) Epicuticular wax In sorghum (Sorghum bicolor L. Moench), drought resistance is a trait that is highly correlated with the thickness of the epicuticular wax layer. Experiments have demonstrated that rice varieties with a thick cuticle layer retain their leaf turgor for longer periods of time after the onset of a water-stress.

(d) Partitioning and Stem Reserve Mobilization

As photosynthesis becomes inhibited by drought, the grain filling process becomes increasingly reliant on stem reserve utilization.72 Numerous studies have reported that stem reserve mobilization capacity is related to yield under water-stress in wheat. In rice, a few studies also indicated that this mechanism maintains grain yield under water stress at the grain filling stage. This drought tolerance mechanism is stimulated by a decrease in gibberellic acid concentration and an increase in abscisic acid concentration.

(e) Manupulation and Stability of flowering processes

(g) Seedling drought traits For emergence from deep sowing (to exploit dry upper soil), this is practiced to help seedlings reach the receding moisture profile, and to avoid high soil surface temperatures which inhibit germination. Screening at these stage provides practical advantages, specially when managing large amount of germplasms.

Crop Ideotype for Drought Tolerance

Usually ideotypes are developed to create an ideal plant variety. The following traits constitutes ideotype of wheat by CIMMYT.

1) Large seed size. Helps emergence, early ground cover, and initial biomass.

2) Long coleoptiles. For emergence from deep sowing

3) Early ground cover. Thinner, wider leaves (i.e., with a relatively low specific leaf weight) and a more prostrate growth habit help to increase ground cover, thus conserving soil moisture and potentially increasing radiation use efficiency.

4) High pre-anthesis biomass.

5) Good capacity for stem reserves and remobilization

6) High spike photosynthetic capacity

7) High RLWC/Gs/CTD during grain filling to indicate ability to extract water

8) Osmotic adjustment

(9) Accumulation of ABA.

The benefit of ABA accumulation under drought has been demonstrated (Innes et al.

1984). It appears to pre-adapt plants to stress by reducing stomatal conductance, rates of cell division, organ size,and increasing development rate. However, high ABA can also result in sterility problems since high ABA levels may abort developing florets

10) Heat Tolerance. The contribution of heat tolerance to performance under moisture stress needs to be quantified, but it is relatively easy to screen for (Reynolds et al. 1998).

11) Leaf anatomy: waxiness, pubescence, rolling, thickness, posture. These traits decrease radiation load to the leaf surface. Benefits include a lower evapotranspiration rate and reduced risk of irreversible photo-inhibition. However, they may also be associated with reduce radiation use efficiency, which would reduce yield under more favorable conditions.

12) High tiller survival.

Comparison of old and new varieties have shown that under drought older varieties over-produce tillers many of which fail to set grain while modern drought tolerant lines produce fewer tillers most of which survive.

13) Stay-green.

The trait may indicate the presence of drought avoidance mechanisms, but probably does not contribute to yield per se if there is no water left in the soil profile by the end of the cycle to support leaf gas exchange. It may be detrimental if it indicates lack of ability to remobilize stem reserves. However, research in sorghum has indicated that stay-green is associated with higher leaf chlorophyll content at all stages of development and both were associated with improved yield and transpiration efficiency under drought.

Combination Phenomics : Overall Health of Crops

The concept of combination phenomics comes from the idea that two or more stresses have common physiological effects or common traits - which are an indicator of overall plant health. Similar analogy in human medical terms is high blood pressure or high body temperature or high white blood cells in body is an indicator of health problems and thus we can select healthy people from unhealthy using such a measure. As both abiotic and abiotic stresses can result in similar physiological consequence, tolerant plant can be separated from sensitive plants. Some imaging or infrared measuring techniques can help to speed the process for breeding process. For example spot blotch intensity and canopy temperature depression can be monitored with canopy temperature depression.

Molecular Breeding for Drought Tolerance

Recent research breakthroughs in biotechnology have revived interest in targeted drought tolerance breeding and use of new genomics tools to enhance crop water productivity. Marker-assisted breeding is revolutionizing the improvement of temperate field crops and will have similar impacts on breeding of tropical crops. Other molecular breeding tool include development of genetically modified crops that can tolerate plant stress. As a complement to the recent rapid progress in genomics, a better understanding of physiological mechanisms of drought response will also contribute to the progress of genetic enhancement of crop drought tolerance. It is now well accepted that the complexity of the drought syndrome can only be tackled with a holistic approach that integrates physiological dissection of crop drought avoidance and tolerance traits using molecular genetic tools such as MAS, microarrays and transgenic crops, with agronomic practices that lead to better conservation and utilization of soil moisture, and better matching of crop genotypes with the environment.

Breeding for Heat Stress Tolerance

Plant breeding is process of development of new cultivars. Plant breeding involves development of varieties for different environmental conditions – some of them are not favorable. Among them, heat stress is one of such factor that reduces the production and quality significantly. So breeding against heat is a very important criterion for breeding for current as well as future environments produced by global climate change (e.g. global warming).

Breeding for Heat Stress Tolerance in Plants

Heat stress due to increased temperature is a very important problem globally. Occasional or prolonged high temperatures cause different morpho-anatomical, physiological and biochemical changes in plants. The ultimate effect is on plant growth as well as development and reduced yield and quality. Breeding for heat stress tolerance can be mitigated by breeding plant varieties that have improved levels of thermo-tolerance using different conventional or advanced genetic tools.

What is Heat Stress Tolerance

Heat stress is defined as increased temperature level sufficient to cause irreversible damage to plant growth and development. Generally a temperature rise, above usually 10 to 15 °C above ambient, can be considered heat shock or heat stress. Heat tolerance is broadly defined as the ability of the plant tolerate heat – means that grow and produce economic yield under high temperatures.

Significance: Current and Future - Global Warming

Heat stress is a serious threat to crop production globally (Hall, 2001, 1992). Global warming is particularly consequence of increased level of green house gases such as CO_2, methane, chlorofluorocarbons and nitrous oxides. The Intergovernmental Panel on Climatic Change (IPCC) has predicted a rise of 0.3 °C per decade (Jones et al., 1999) reaching to approximately 1 and 3 °C above the present value by 2025 and 2100 AD, respectively.

Physiological Consequence of Heat Stress

At very high temperatures cause severe cellular injury and cell death may occur within short time, thus leading to a catastrophic collapse of cellular organization (Schoffl et al., 1999). However, under moderately high temperatures, the injury can only occur after longer exposure to such a temperature however the plant efficiency can be severely affected. High temperature directly affect injuries such as protein denaturation and aggregation, and increased fluidity of membrane lipids. Other indirect or slower heat injuries involve inactivation of enzymes in chloroplast and mitochondria, protein degradation, inhibition of protein synthesis, and loss of membrane integrity. Heat stress associated injuries ultimately lead to starvation, inhibition of growth, reduced ion flux, production of toxic compounds and production of reactive oxygen species (ROS). Immediately after exposure to high temperature stress-related proteins are expressed as stress defense strategy of the cell. Expression of heat shock proteins (HSPs), protein with 10 to 200 kDa, is supposed to be involved in signal transduction during heat stress. In many species it has been demonstrated that HSPs results in improved physiological phenomena such as photosynthesis, assimilate partitioning, water and nutrient use efficiency, and membrane stability.

Studies have found tremendous variation within and between species, thus this will help to breed heat tolerance for future environment. Some of attempts to develop heat-tolerant genotypes are successful. (Ehlers and Hall, 1998; Camejo et al., 2005)

Traits Associated with Heat Stress Tolerance

Different physiological mechanisms may contribute to heat tolerance in the field—for example, heat tolerant metabolism as indicated by higher photosynthetic rates, stay-green, and membrane thermo-stability, or heat avoidance as indicated by canopy temperature depression. Several physiological and morphological traits have been evaluated for heat tolerance - Canopy temperature, leaf chlorophyll, stay green, leaf conductance, spike number, biomass, and flowering date.

(a) Canopy Temperature Depression (CTD)

CTD has shown clear association with yield in warm environments shows it association with heat stress tolerance. CTD shows high genetic correlation with yield and high values of proportion of direct response to selection (Reynolds et al., 1998) indicating that the trait is heritable and therefore amenable to early generation selection. Since an integrated CTD value can be measured almost instantaneously on scores of plants in a small breeding plot (thus reducing error normally associated with traits measured on individual plants), work has been conducted to evaluate its potential as an indirect selection criterion for genetic gains in yield. CTD is affected by many physiological factors, which makes it a powerful.

(b) Stomatal Conductance

Canopy temperature depression is highly suitable for selecting physiologically superior lines in warm, low relative humidity environments where high evaporative demand leads to leaf cooling of up to 10 °C below ambient temperatures. This permits differences among genotypes to be detected relatively easily using infrared thermometry. However, such differences cannot be detected in high relative humidity environments because the effect of evaporative cooling of leaves is negligible. Nonetheless, leaves maintain their stomata open to permit the uptake of CO_2, and differences in the rate of CO_2 fixation may lead to differences in leaf conductance that can be measured using a porometer. Porometry can be used to screen individual plants. The heritability of stomatal conductance is reasonably high, with reported values typically in the range of 0.5 to 0.8. Plants can be assessed for leaf conductance using a viscous flow porometer that is available on the market (Thermoline and CSIRO, Australia). This instrument can give a relative measure of stomatal conductance in a few seconds, making it possible to identify physiologically superior genotypes from within bulks.

(C) Membrane Thermostability

Although resistance to high temperatures involves several complex tolerance and avoidance mechanisms, the membrane is thought to be a site of primary physiological injury by heat, and measurement of solute leakage from tissue can be used to estimate damage to membranes. Since membrane thermostability is reasonably heritable (Fokar et al., 1998) and shows high genetic correlation with yield.

(D) Chlorophyll Fluorescence

Chlorophyll fluorescence, an indication of the fate of excitation energy in the photosynthetic apparatus, has been used indicator for heat stress tolerance.

(E) Chlorophyll Content and Stay Green

Chlorophyll content and stay green traits have been found to be associated with heat stress tolerance.,. Xu et al. (2000) identified three QTLs for chlorophyll content (Chl1, Chl2, and Chl3) (coincided with three stay-green QTL regions (Stg1, Stg2, and Stg3)) were identified in Sorghum. The Stg1 and Stg2 regions also contain the genes for key photosynthetic enzymes, heat shock proteins, and an abscisic acid (ABA) responsive gene.

(F) Photosynthesis

Decliend photosynthesis is suggested as measure of heat stress sensitivity in plants.

(G) Stem Reserve Remobilization'

Combination Breeding and Physiological Breeding

The physiological-trait-based breeding approach has merit over breeding for yield *per se* because it increases the probability of crosses resulting in additive gene action. The concept of combination phenomics comes from the idea that two or more stress have common physiological effect or common traits - which is an indicator of overall plant health. Similar analogy in human medical terms is high blood pressure or high body temperature or high white blood cells in body is an indicator of health problems and thus we can select healthy people from unhealthy using such a measure. As both abiotic and abiotic stresses can result in similar physiological consequence, tolerant plant can be separated from sensitive plants. Some imaging or infrared measuring techniques can help to speed the process for breeding process. For example, spot blotch intensity and canopy temperature depression can be monitored with canopy temperature depression.

References

- Alberts, Bruce; et al. (2002). Molecular Biology of the Cell. New York: Garland Science. p. G:35. ISBN 978-0-8153-4072-0.
- Srivastava, Sheela (2013). Genetics of Bacteria (PDF). India: Springer-Verlag. doi:10.1007/978-81-322-1090-0. ISBN 978-81-322-1089-4.
- Mägdefrau, Karl (1992). Geschichte der Botanik [History of Botany] (2 ed.). Jena: Gustav Fischer Verlag. ISBN 3-437-20489-0.
- Cristianini, N. and Hahn, M. Introduction to Computational Genomics, Cambridge University Press, 2006. (ISBN 9780521671910 | ISBN 0-521-67191-4)
- Panesar, Pamit et al (2010) "Enzymes in Food Processing: Fundamentals and Potential Applications", Chapter 10, I K International Publishing House, ISBN 978-9380026336
- Shiv Kant Prasad, Ajay Dash (2008). Modern Concepts in Nanotechnology, Volume 5. Discovery Publishing House. ISBN 9788183562966.
- Head, Graham; Hull, Roger H; Tzotzos, George T. (2009). Genetically Modified Plants: Assessing Safety and Managing Risk. London: Academic Pr. p. 244. ISBN 0-12-374106-8.
- Avise, John C. (2004). The hope, hype & reality of genetic engineering: remarkable stories from agriculture, industry, medicine, and the environment. Oxford University Press US. p. 22. ISBN 978-0-19-516950-8.
- Pollack, Andrew (19 November 2015). "Genetically Engineered Salmon Approved for Consumption". The New York Times. Retrieved 21 April 2016.
- Angulo, E.; Cooke, B. (2002). "First synthesize new viruses then regulate their release? The case of the wild rabbit". Molecular Ecology. 11: 2703–9. doi:10.1046/j.1365-294X.2002.01635.x. PMID 12453252. Retrieved 16 May 2016.
- Thomas; et al. (25 September 2013). "Ecology: Gene tweaking for conservation". Nature. 501: 485–6. doi:10.1038/501485a. PMID 24073449. Retrieved 16 May 2016.
- "The Morphology of Eukaryotic Cells: Shape, Number and Size". YourArticleLibrary.com: The Next Generation Library. Retrieved 2015-11-22.

4

Issues and Concerns of Plant Breeding

Plant breeding comes with issues of concern, particularly in regard to food crop. The aim of this chapter is to provide an in-depth understanding on the issues and concerns of plant breeding. Nutrition and plant breeder rights are some of the significant and important topics related to plant breeding. The following chapter unfolds its crucial aspects in a critical yet systematic manner.

Nutrition

Nutrition is the science that interprets the interaction of nutrients and other substances in food in relation to maintenance, growth, reproduction, health and disease of an organism. It includes food intake, absorption, assimilation, biosynthesis, catabolism and excretion.

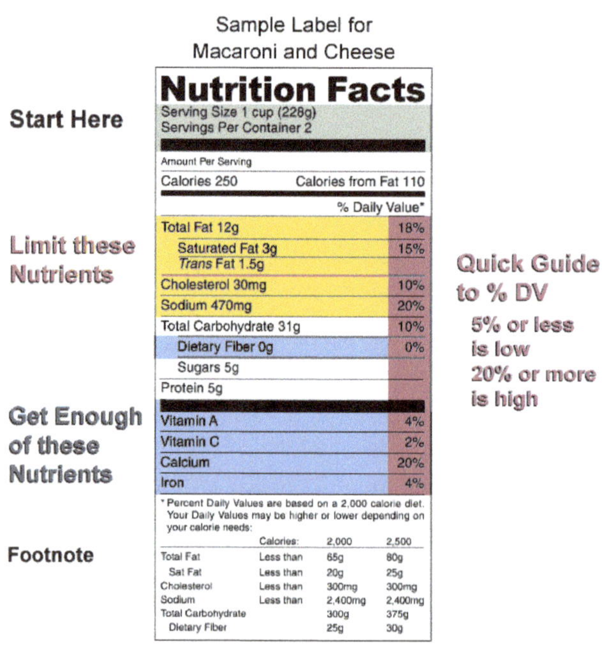

The "Nutrition Facts" table displayed on packaged food labels in the United States indicates the amounts of nutrients recommended by the US Food and Drug Administration to limit or consume according to the Daily Value

The diet of an organism is what it eats, which is largely determined by the availability, the processing and palatability of foods. A healthy diet includes preparation of food and storage methods that preserve nutrients from oxidation, heat or leaching, and that reduce risk of foodborne illness.

Registered dietitian nutritionists (RDs or RDNs) are health professionals qualified to provide safe, evidence-based dietary advice which includes a review of what is eaten, a thorough review of nutritional health, and a personalized nutritional treatment plan. They also provide preventive and therapeutic programs at work places, schools and similar institutions. Certified Clinical Nutritionists or CCNs, are trained health professionals who also offer dietary advice on the role of nutrition in chronic disease, including possible prevention or remediation by addressing nutritional deficiencies before resorting to drugs. Government regulation especially in terms of licensing, is currently less universal for the CCN than that of RD or RDN. Another advanced Nutrition Professional is a Certified Nutrition Specialist or CNS. These Board Certified Nutritionists typically specialize in obesity and chronic disease. In order to become board certified, potential CNS candidate must pass an examination, much like Registered Dieticians. This exam covers specific domains within the health sphere including; Clinical Intervention and Human Health.

A poor diet may cause health problems, causing deficiency diseases such as blindness, anemia, scurvy, preterm birth, stillbirth and cretinism; health-threatening conditions like obesity and metabolic syndrome; and such common chronic systemic diseases as cardiovascular disease, diabetes, and osteoporosis. A poor diet can cause the wasting of kwashiorkor in acute cases, and the stunting of marasmus in chronic cases of malnutrition.

History

Antiquity

Hippocrates lived about 400 BC, and Galen and the understanding of nutrition followed him for centuries.

The first recorded dietary advice, carved into a Babylonian stone tablet in about 2500 BC, cautioned those with pain inside to avoid eating onions for three days. Scurvy, later found to be a vitamin C deficiency, was first described in 1500 BC in the Ebers Papyrus.

According to Walter Gratzer, the study of nutrition probably began during the 6th century BC. In China, the concept of *Qi* developed, a spirit or "wind" similar to what Western Europeans later called *pneuma*. Food was classified into "hot" (for example, meats, blood, ginger, and hot spices) and "cold" (green vegetables) in China, India, Malaya, and Persia. *Humours* developed perhaps first in China alongside *qi*. Ho the Physician concluded that diseases are caused by deficiencies of elements (Wu Xing: fire, water, earth, wood, and metal), and he classified diseases as well as prescribed diets. About the same time in Italy, Alcmaeon of Croton (a Greek) wrote of the importance of equilibrium between what goes in and what goes out, and warned that imbalance would result in disease marked by obesity or emaciation.

The first recorded nutritional experiment with human subjects is found in the Bible's Book of Daniel. Daniel and his friends were captured by the king of Babylon during an invasion of Israel. Selected as court servants, they were to share in the king's fine foods and wine. But they objected, preferring vegetables (pulses) and water in accordance with their Jewish dietary restrictions. The king's chief steward reluctantly agreed to a trial. Daniel and his friends received their diet for ten days and were then compared to the king's men. Appearing healthier, they were allowed to continue with their diet.

Around 475 BC, Anaxagoras stated that food is absorbed by the human body and, therefore, contains "homeomerics" (generative components), suggesting the existence of nutrients. Around 400 BC, Hippocrates, who recognized and was concerned with obesity, which may have been common in southern Europe at the time, said, "Let food be your medicine and medicine be your food." The works that are still attributed to him, *Corpus Hippocraticum*, called for moderation and emphasized exercise.

Followed for a millennium and a half, Galen (1st century) created the first coherent (although mistaken) theory of nutrition.

Salt, pepper and other spices were prescribed for various ailments in various preparations for example mixed with vinegar. In the 2nd century BC, Cato the Elder believed that cabbage (or the urine of cabbage-eaters) could cure digestive diseases, ulcers, warts, and intoxication. Living about the turn of the millennium, Aulus Celsus, an ancient Roman doctor, believed in "strong" and "weak" foods (bread for example was strong, as were older animals and vegetables).

Galen to Lind

One mustn't overlook the doctrines of Galen: In use from his life in the 1st century AD until the 17th century, it was heresy to disagree with him for 1500 years. Galen was physician to gladiators in Pergamon, and in Rome, physician to Marcus Aurelius and the three emperors who succeeded him. Most of Galen's teachings were gathered and enhanced in the late 11th century by Benedictine monks at the School of Salerno in *Regimen sanitatis Salernitanum*, which still had users in the 17th century. Galen believed in the bodily *humours* of Hippocrates, and he taught that *pneuma* is the source of life. Four elements (earth, air, fire and water) combine into "complexion", which combines into states (the four temperaments: sanguine, phlegmatic, choleric, and melancholic). The states are made up of pairs of attributes (hot and moist, cold and moist, hot and dry, and cold and dry), which are made of four humours: blood, phlegm, green (or yellow) bile, and black bile (the bodily form of the elements). Galen thought that for a person to have gout, kidney stones, or arthritis was scandalous, which Gratzer likens to Samuel Butler's *Erehwon* (1872) where sickness is a crime.

James Lind conducted in 1747 the first controlled clinical trial in modern times, and in 1753 published *Treatise on Scurvy*.

In the 1500s, Paracelsus was probably the first to criticize Galen publicly. Also in the 16th century, scientist and artist Leonardo da Vinci compared metabolism to a burning candle. Leonardo did not publish his works on this subject, but he was not afraid of thinking for himself and he definitely disagreed with Galen. Ultimately, 16th century works of Andreas Vesalius, sometimes called the father of modern medicine, overturned Galen's ideas. He was followed by piercing thought amalgamated with the era's mysticism and religion sometimes fueled by the mechanics of Newton and Galileo. Jan Baptist van Helmont, who discovered several gases such as carbon dioxide, performed the first quantitative experiment. Robert Boyle advanced chemistry. Sanctorius measured body weight. Physician Herman Boerhaave modeled the digestive process. Physiologist Albrecht von Haller worked out the difference between nerves and muscles.

Sometimes overlooked during his life, James Lind, a physician in the British navy, performed the first scientific nutrition experiment in 1747. Lind discovered that lime juice saved sailors that had been at sea for years from scurvy, a deadly and painful bleeding disorder. Between 1500 and 1800, an estimated two million sailors had died of scurvy. The discovery was ignored for forty years, after which British sailors became known as "limeys." The essential vitamin C within citrus fruits would not be identified by scientists until 1932.

Lavoisier and Modern Science

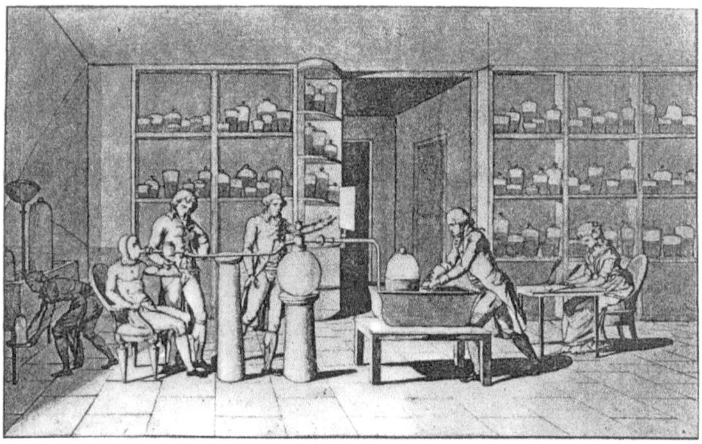

By containing his assistant, Armand Seguin, inside a rubber suit fitted with a tube sealed to his mouth with putty, Antoine Lavoisier first measured basal metabolic rate. Drawing by Madame Lavoisier (seated at right).

Around 1770, Antoine Lavoisier discovered the details of metabolism, demonstrating that the oxidation of food is the source of body heat. Called the most fundamental chemical discovery of the 18th century, Lavoisier discovered the principle of conservation of mass. His ideas made the phlogiston theory of combustion obsolete.

In 1790, George Fordyce recognized calcium as necessary for the survival of fowl. In the early 19th century, the elements carbon, nitrogen, hydrogen, and oxygen were recognized as the primary components of food, and methods to measure their proportions were developed.

In 1816, François Magendie discovered that dogs fed only carbohydrates (sugar), fat (olive oil), and water died evidently of starvation, but dogs also fed protein survived, identifying protein as an essential dietary component. William Prout in 1827 was the first person to divide foods into carbohydrates, fat, and protein. During the 19th century, Jean-Baptiste Dumas and Justus von Liebig quarrelled over their shared belief that animals get their protein directly from plants (animal and plant protein are the same and that humans do not create organic compounds). With a reputation as the leading organic chemist of his day but with no credentials in animal physiology, Liebig grew rich making food extracts like beef bouillon and infant formula that were later found to be of questionable nutritious value. In the 1860s, Claude Bernard discovered that body fat can be synthesized from carbohydrate and protein, showing that the energy in blood glucose can be stored as fat or as glycogen.

In the early 1880s, Kanehiro Takaki observed that Japanese sailors (whose diets consisted almost entirely of white rice) developed beriberi (or endemic neuritis, a disease causing heart problems and paralysis), but British sailors and Japanese naval officers did not. Adding various types of vegetables and meats to the diets of Japanese sailors prevented the disease, (not because of the increased protein as Takaki supposed but because it introduced a few parts per million of thiamine to the diet, later understood as a cure).

In 1896, Eugen Baumann observed iodine in thyroid glands. In 1897, Christiaan Eijkman worked with natives of Java, who also suffered from beriberi. Eijkman observed that chickens fed the native diet of white rice developed the symptoms of beriberi but remained healthy when fed unprocessed brown rice with the outer bran intact. Eijkman cured the natives by feeding them brown rice, discovering that food can cure disease. Over two decades later, nutritionists learned that the outer rice bran contains vitamin B1, also known as thiamine.

From 1900 to the Present

Frederick Hopkins discovered vitamins, for which he shared a Nobel prize with Eijkman.

In the early 20th century, Carl von Voit and Max Rubner independently measured caloric energy expenditure in different species of animals, applying principles of physics in nutrition. In 1906, Edith G. Willcock and Frederick Hopkins showed that the amino acid tryptophan aids the well-being of mice but it did not assure their growth. In the middle of twelve years of attempts to isolate them, Hopkins said in a 1906 lecture that "unsuspected dietetic factors," other than calories, protein, and minerals, are needed to prevent deficiency diseases. In 1907, Stephen M. Babcock and Edwin B. Hart conducted the single-grain experiment, which took nearly four years to complete.

Vitamin	Year Isolated
Thiamin	1926
Vitamin C	1926
Vitamin A	1939
Vitamin D	1931
Vitamin E	1936
Niacin	1937
Biotin	1939
Vitamin K	1939
Pantothenic acid	1939
Folate	1939
Riboflavin	1933
Vitamin B6	1936

Oxford University closed down its nutrition department after World War II because the subject seemed to have been completed between 1912 and 1944.

In 1912, Casimir Funk coined the term vitamin, a vital factor in the diet, from the words "vital" and "amine," because these unknown substances preventing scurvy, beriberi, and pellagra, were thought then to be derived from ammonia. The vitamins were studied in the first half of the 20th century.

In 1913, Elmer McCollum and Marguerite Davis discovered the first vitamin, fat-soluble vitamin A, then water-soluble vitamin B (in 1915; now known to be a complex of several water-soluble vitamins) and named vitamin C as the then-unknown substance preventing scurvy. Lafayette Mendel and Thomas Osborne also performed pioneering work on vitamins A and B. In 1919, Sir Edward Mellanby incorrectly identified rickets as a vitamin A deficiency because he could cure it in dogs with cod liver oil. In 1922, McCollum destroyed the vitamin A in cod liver oil, but found that it still cured rickets. Also in 1922, H.M. Evans and L.S. Bishop discover vitamin E as essential for rat pregnancy, originally calling it "food factor X" until 1925.

In 1925, Hart discovered that trace amounts of copper are necessary for iron absorption. In 1927, Adolf Otto Reinhold Windaus synthesized vitamin D, and was awarded the Nobel Prize in Chemistry in 1928. In 1928, Albert Szent-Györgyi isolated ascorbic acid, and in 1932 proved that it is vitamin C by preventing scurvy. In 1935, he synthesized it, and in 1937, he won a Nobel Prize for his efforts. Szent-Györgyi concurrently elucidated much of the citric acid cycle.

In the 1930s, William Cumming Rose identified essential amino acids, necessary protein components that the body cannot synthesize. In 1935, Underwood and Marston independently discovered the necessity of cobalt. In 1936, Eugene Floyd DuBois showed that work and school performance are related to caloric intake. In 1938, Erhard Fernholz discovered the chemical structure of vitamin E and then he tragically disappeared. It was synthesised the same year by Paul Karrer.

In 1940, rationing in the United Kingdom during and after World War II took place according to nutritional principles drawn up by Elsie Widdowson and others. In 1941, the first Recommended Dietary Allowances (RDAs) were established by the National Research Council.

In 1992, The U.S. Department of Agriculture introduced the Food Guide Pyramid. In 2002, a Natural Justice study showed a relation between nutrition and violent behavior. In 2005, one inconclusive study found that obesity could be caused by adenovirus in addition to bad nutrition.

World leaders are looking at alternatives like genetically modified foods to tackle the problem of world hunger and food shortages.

Nutrients

The list of nutrients that people are known to require is, in the words of Marion Nestle, "almost certainly incomplete". As of 2014, nutrients are thought to be of two types: macro-nutrients which are needed in relatively large amounts, and micronutrients which are needed in smaller quantities. A type of carbohydrate, dietary fiber, i.e. non-digestible material such as cellulose, is required, for both mechanical and biochemical reasons, although the exact reasons remain unclear. Other micronutrients include antioxidants and phytochemicals, which are said to influence (or protect) some body systems. Their necessity is not as well established as in the case of, for instance, vitamins.

Most foods contain a mix of some or all of the nutrient types, together with other substances, such as toxins of various sorts. Some nutrients can be stored internally (e.g., the fat-soluble vitamins), while others are required more or less continuously. Poor health can be caused by a lack of required nutrients or, in extreme cases, too much of a required nutrient. For example, both salt and water (both absolutely required) will cause illness or even death in excessive amounts.

Macronutrients

The macronutrients are carbohydrates, fiber, fats, protein, and water. The macronutrients (excluding fiber and water) provide structural material (amino acids from which proteins are built, and lipids from which cell membranes and some signaling molecules are built) and energy. Some of the structural material can be used to generate energy internally, and in either case it is measured in Joules or kilocalories (often called "Calories" and written with a capital *C* to distinguish them from little 'c' calories). Carbohydrates and proteins provide 17 kJ approximately (4 kcal) of energy per gram, while fats provide 37 kJ (9 kcal) per gram, though the net energy from either depends on such factors as absorption and digestive effort, which vary substantially from instance to instance. Vitamins, minerals, fiber, and water do not provide energy, but are required for other reasons.

Molecules of carbohydrates and fats consist of carbon, hydrogen, and oxygen atoms. Carbohydrates range from simple monosaccharides (glucose, fructose, galactose) to complex polysaccharides (starch). Fats are triglycerides, made of assorted fatty acid monomers bound to a glycerol backbone. Some fatty acids, but not all, are essential in the diet: they cannot be synthesized in the body. Protein molecules contain nitrogen atoms in addition to carbon, oxygen, and hydrogen. The fundamental components of protein are nitrogen-containing amino acids, some of which are essential in the sense that humans cannot make them internally. Some of the amino acids are convertible (with the expenditure of energy) to glucose and can be used for energy production, just as ordinary glucose, in a process known as gluconeogenesis. By breaking down existing protein, the carbon skeleton of the various amino acids can be metabolized to intermediates in cellular respiration; the remaining ammonia is discarded primarily as urea in urine. This occurs normally only during prolonged starvation.

Carbohydrates

Carbohydrates may be classified as monosaccharides, disaccharides, or polysaccharides depending on the number of monomer (sugar) units they contain. They constitute a large part of foods such as rice, noodles, bread, and other grain-based products. Monosaccharides, disaccharides, and polysaccharides contain one, two, and three or more sugar units, respectively. Polysaccharides are often referred to as *complex* carbohydrates because they are typically long, multiple branched chains of sugar units.

Traditionally, simple carbohydrates are believed to be absorbed quickly, and therefore to raise blood-glucose levels more rapidly than complex carbohydrates. This, however, is not accurate. Some simple carbohydrates (e.g., fructose) follow different metabolic pathways (e.g., fructolysis) that result in only a partial catabolism to glucose, while, in essence, many complex carbohydrates may be digested at the same rate as simple carbohydrates. Glucose stimulates the production of insulin through food entering the bloodstream, which is grasped by the beta cells in the pancreas.

Fiber

Dietary fiber is a carbohydrate that is incompletely absorbed in humans and in some animals. Like all carbohydrates, when it is metabolized it can produce four Calories (kilocalories) of energy per gram. However, in most circumstances it accounts for less than that because of its limited absorption and digestibility. Dietary fiber consists mainly of cellulose, a large carbohydrate polymer which is indigestible as humans do not have the required enzymes to disassemble it. There are two subcategories: soluble and insoluble fiber. Whole grains, fruits (especially plums, prunes, and figs), and vegetables are good sources of dietary fiber. There are many health benefits of a high-fiber diet. Dietary fiber helps reduce the chance of gastrointestinal problems such as constipation and diarrhea by increasing the weight and size of stool and softening it. Insoluble fiber, found in whole wheat flour, nuts and vegetables, especially stimulates peristalsis – the rhythmic muscular contractions of the intestines, which move digesta along the digestive tract. Soluble fiber, found in oats, peas, beans, and many fruits, dissolves in water in the intestinal tract to produce a gel that slows the movement of food through the intestines. This may help lower blood glucose levels because it can slow the absorption of sugar. Additionally, fiber, perhaps especially that from whole grains, is thought to possibly help lessen insulin spikes, and therefore reduce the risk of type 2 diabetes. The link between increased fiber consumption and a decreased risk of colorectal cancer is still uncertain.

Fat

A molecule of dietary fat typically consists of several fatty acids (containing long chains of carbon and hydrogen atoms), bonded to a glycerol. They are typically found as triglycerides (three fatty acids attached to one glycerol backbone). Fats may be classified as saturated or unsaturated depending on the detailed structure of the fatty acids involved. Saturated fats have all of the carbon atoms in their fatty acid chains bonded to hydrogen atoms, whereas unsaturated fats have some of these carbon atoms double-bonded, so their molecules have relatively fewer hydrogen atoms than a saturated fatty acid of the same length. Unsaturated fats may be further classified as monounsaturated (one double-bond) or polyunsaturated (many double-bonds). Furthermore, depending on the location of the double-bond in the fatty acid chain, unsaturated fatty acids are classified as omega-3 or omega-6 fatty acids. Trans fats are a type of unsaturated fat with *trans*-isomer bonds; these are rare in nature and in foods from natural sources; they are typically created in an industrial process called (partial) hydrogenation. There are nine kilocalories in each gram of fat. Fatty acids such as conjugated linoleic acid, catalpic acid, eleostearic acid and punicic acid, in addition to providing energy, represent potent immune modulatory molecules.

Saturated fats (typically from animal sources) have been a staple in many world cultures for millennia. Unsaturated fats (e. g., vegetable oil) are considered healthier, while trans fats are to be avoided. Saturated and some trans fats are typically solid at room temperature (such as butter or lard), while unsaturated fats are typically liquids (such as olive oil or flaxseed oil). Trans fats are very rare in nature, and have been shown to be highly detrimental to human health, but have properties useful in the food processing industry, such as rancidity resistance.

Essential Fatty Acids

Most fatty acids are non-essential, meaning the body can produce them as needed, generally from other fatty acids and always by expending energy to do so. However, in humans, at least two fatty acids are essential and must be included in the diet. An appropriate balance of essential fatty acids—omega-3 and omega-6 fatty acids—seems also important for health, although definitive experimental demonstration has been elusive. Both of these "omega" long-chain polyunsaturated fatty acids are substrates for a class of eicosanoids known as prostaglandins, which have roles throughout the human body. They are hormones, in some respects. The omega-3 eicosapentaenoic acid (EPA), which can be made in the human body from the omega-3 essential fatty acid alpha-linolenic acid (ALA), or taken in through marine food sources, serves as a building block for series 3 prostaglandins (e.g., weakly inflammatory PGE3). The omega-6 dihomo-gamma-linolenic acid (DGLA) serves as a building block for series 1 prostaglandins (e.g. anti-inflammatory PGE1), whereas arachidonic acid (AA) serves as a building block for series 2 prostaglandins (e.g. pro-inflammatory PGE 2). Both DGLA and AA can be made from the omega-6 linoleic acid (LA) in the human body, or can be taken in directly through food. An appropriately balanced intake of omega-3 and omega-6 partly determines the relative production of different prostaglandins, which is one reason why a balance between omega-3 and omega-6 is believed important for cardiovascular health. In industrialized societies, people typically consume large amounts of processed vegetable oils, which have reduced amounts of the essential fatty acids along with too much of omega-6 fatty acids relative to omega-3 fatty acids.

The conversion rate of omega-6 DGLA to AA largely determines the production of the prostaglandins PGE1 and PGE2. Omega-3 EPA prevents AA from being released from membranes, thereby skewing prostaglandin balance away from pro-inflammatory PGE2 (made from AA) toward anti-inflammatory PGE1 (made from DGLA). Moreover, the conversion (desaturation) of DGLA to AA is controlled by the enzyme delta-5-desaturase, which in turn is controlled by hormones such as insulin (up-regulation) and glucagon (down-regulation). The amount and type of carbohydrates consumed, along with some types of amino acid, can influence processes involving insulin, glucagon, and other hormones; therefore, the ratio of omega-3 versus omega-6 has wide effects on general health, and specific effects on immune function and inflammation, and mitosis (i.e., cell division).

Protein

Proteins are structural materials in much of the animal body (e.g. muscles, skin, and hair). They also form the enzymes that control chemical reactions throughout the body. Each protein molecule is composed of amino acids, which are characterized by inclusion of nitrogen and sometimes sulphur (these components are responsible for the distinctive smell of burning protein, such as the keratin in hair). The body requires amino acids to produce new proteins (protein retention) and to replace damaged proteins (maintenance). As there is no protein or amino acid storage provision, amino acids must be present in the diet. Excess amino acids are discarded, typically in the urine. For all animals, some amino acids are *essential* (an animal cannot produce them internally) and some are *non-essential* (the animal can produce them from other nitrogen-containing compounds). About twenty amino acids are found in the human body, and about ten of these are essential and, therefore, must be included in the diet. A diet that contains adequate amounts of amino acids (especially those that are essential) is particularly important in some situations: during early development and maturation, pregnancy, lactation, or injury (a burn, for instance). A *complete* protein source contains all the essential amino acids; an *incomplete* protein source lacks one or more of the essential amino acids.

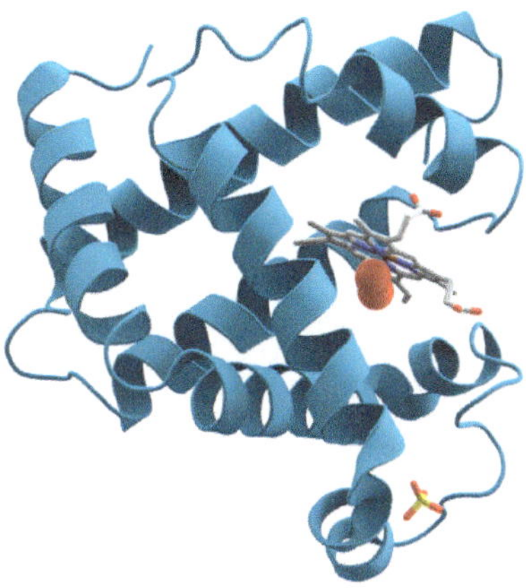

Proteins are chains of amino acids found in most nutritional foods.

It is possible with protein combinations of two incomplete protein sources (e.g., rice and beans) to make a complete protein source, and characteristic combinations are the basis of distinct cultural cooking traditions. However, complementary sources of protein do not need to be eaten at the same meal to be used together by the body. Excess amino acids from protein can be converted into glucose and used for fuel through a process called gluconeogenesis. The amino acids remaining after such conversion are discarded.

Water

Water is excreted from the body in multiple forms; including urine and feces, sweating, and by water vapour in the exhaled breath. Therefore, it is necessary to adequately rehydrate to replace lost fluids.

A manual water pump in China

Early recommendations for the quantity of water required for maintenance of good health suggested that 6–8 glasses of water daily is the minimum to maintain proper hydration. However the notion that a person should consume eight glasses of water per day cannot be traced to a credible scientific source. The original water intake recommendation in 1945 by the Food and Nutrition Board of the National Research Council read: "An ordinary standard for diverse persons is 1 milliliter for each calorie of food. Most of this quantity is contained in prepared foods." More recent comparisons of well-known recommendations on fluid intake have revealed large discrepancies in the volumes of water we need to consume for good health. Therefore, to help standardize guidelines, recommendations for water consumption are included in two recent European Food Safety Authority (EFSA) documents (2010): (i) Food-based dietary guidelines and (ii) Dietary reference values for water or adequate daily intakes (ADI). These specifications were provided by calculating adequate intakes from measured intakes in populations of individuals with "desirable osmolarity values of urine and desirable water volumes per energy unit consumed." For healthful hydration, the current EFSA guidelines recommend total water intakes of 2.0 L/day for adult females and 2.5 L/day for adult males. These reference values include water from drinking water, other beverages, and from food. About 80% of our daily water requirement comes from the beverages we drink, with the remaining 20% coming from food. Water content varies depending on the type of food consumed, with fruit and vegetables containing more than cereals, for example. These values are estimated using country-specific food balance sheets published by the Food and Agriculture Organisation of the United Nations. Other guidelines for nutrition also have implications for the beverages we consume for healthy hydration- for example, the World Health Organization (WHO) recommend that added sugars should represent no more than 10% of total energy intake.

The EFSA panel also determined intakes for different populations. Recommended intake volumes in the elderly are the same as for adults as despite lower energy consumption, the water requirement of this group is increased due to a reduction in renal concentrating capacity. Pregnant and breastfeeding women require additional fluids to stay hydrated. The EFSA panel proposes that pregnant women should consume the same volume of water as non-pregnant women, plus an increase in proportion to the higher energy requirement, equal to 300 mL/day. To compensate for additional fluid output, breastfeeding women require an additional 700 mL/day above the recommended intake values for non-lactating women.

For those who have healthy kidneys, it is somewhat difficult to drink too much water, but (especially in warm humid weather and while exercising) it is dangerous to drink too little. While overhydration is much less common than dehydration, it is also possible to drink far more water than necessary, which can result in water intoxication, a serious and potentially fatal condition. In particular, large amounts of de-ionized water are dangerous.

Micronutrients

The micronutrients are minerals, vitamins, and others.

Minerals

Dietary minerals are inorganic chemical elements required by living organisms, other than the four elements carbon, hydrogen, nitrogen, and oxygen that are present in nearly all organic molecules. The term "mineral" is archaic, since the intent is to describe simply the less common elements in the diet. Some are heavier than the four just mentioned, including several metals, which often occur as ions in the body. Some dietitians recommend that these be supplied from foods in which they occur naturally, or at least as complex compounds, or sometimes even from natural inorganic sources (such as calcium carbonate from ground oyster shells). Some minerals are absorbed much more readily in the ionic forms found in such sources. On the other hand, minerals are often artificially added to the diet as supplements; the most famous is likely iodine in iodized salt which prevents goiter.

Macrominerals

Many elements are essential in relative quantity; they are usually called "bulk minerals". Some are structural, but many play a role as electrolytes. Elements with recommended dietary allowance (RDA) greater than 150 mg/day are, in alphabetical order (with informal or folk-medicine perspectives in parentheses):

- Calcium, a common electrolyte, but also needed structurally (for muscle and digestive system health, bone strength, some forms neutralize acidity, may help clear toxins, provides signaling ions for nerve and membrane functions)
- Chlorine as chloride ions; very common electrolyte; see sodium, below
- Magnesium, required for processing ATP and related reactions (builds bone,

causes strong peristalsis, increases flexibility, increases alkalinity)

- Phosphorus, required component of bones; essential for energy processing
- Potassium, a very common electrolyte (heart and nerve health)
- Sodium, a very common electrolyte; in general not found in dietary supplements, despite being needed in large quantities, because the ion is very common in food: typically as sodium chloride, or common salt. Excessive sodium consumption can deplete calcium and magnesium, leading to high blood pressure and osteoporosis.
- Sulfur, for three essential amino acids and therefore many proteins (skin, hair, nails, liver, and pancreas). Sulfur is not consumed alone, but in the form of sulfur-containing amino acids

Trace Minerals

Many elements are required in trace amounts, usually because they play a catalytic role in enzymes. Some trace mineral elements (RDA < 200 mg/day) are, in alphabetical order:

- Cobalt required for biosynthesis of vitamin B12 family of coenzymes. Animals cannot biosynthesize B12, and must obtain this cobalt-containing vitamin in their diet.
- Copper required component of many redox enzymes, including cytochrome c oxidase
- Chromium required for sugar metabolism
- Iodine required not only for the biosynthesis of thyroxine but also — it is presumed — for other important organs as breast, stomach, salivary glands, thymus, etc. For this reason iodine is needed in larger quanti-ties than others in this list, and sometimes classified with the macrominerals.
- Iron required for many enzymes, and for hemoglobin and some other proteins
- Manganese (processing of oxygen)
- Molybdenum required for xanthine oxidase and related oxidases
- Selenium required for peroxidase (antioxidant proteins)
- Zinc required for several enzymes such as carboxypeptidase, liver alcohol dehydrogenase, and carbonic anhydrase

Vitamins

As with the minerals discussed above, some vitamins are recognized as organic essential nutrients, necessary in the diet for good health. (Vitamin D is the exception: it can be synthesized in the skin, in the presence of UVB radiation.) Certain vitamin-like compounds that are recommended in the diet, such as carnitine, are thought useful for survival and health, but these are not "essential" dietary nutrients because the human body has some capacity to produce them from other compounds. Moreover, thousands of different phytochemicals have recently been discovered in food (particularly in fresh vegetables), which may have desirable properties including antioxidant activity however, experimental demonstration has been suggestive but inconclusive. Other essential nutrients that are not classified as vitamins include essential amino acids, choline, essential fatty acids, and the minerals discussed in the preceding section.

Vitamin deficiencies may result in disease conditions, including goitre, scurvy, osteoporosis, impaired immune system, disorders of cell metabolism, certain forms of cancer, symptoms of premature aging, and poor psychological health (including eating disorders), among many others. Excess levels of some vitamins are also dangerous to health (notably vitamin A). Deficient or excess levels of minerals can also have serious health consequences.

Healthy Diets

Whole Plant Food Diet

T. Colin Campbell is among the scientists who advocate a plant-based diet

Heart disease, cancer, obesity, and diabetes are commonly called "Western" diseases because these maladies were once rarely seen in developing countries. An international study in China found some regions had virtually no cancer or heart disease, while in other areas they reflected "up to a 100-fold increase" coincident with shifts from diets that were found to be entirely plant-based to heavily animal-based, respectively. In contrast, diseases of affluence like cancer and heart disease are common throughout the developed world, including the United States. Adjusted for age and exercise, large regional clusters of people in China rarely suffered from these "Western" diseases possibly because their diets are rich in vegetables, fruits, and whole grains, and have little dairy and meat products. Some studies show these to be, in high quantities, possible causes of some cancers. There are arguments for and against this controversial issue.

The United Healthcare/Pacificare nutrition guideline recommends a whole plant food diet, and recommends using protein only as a condiment with meals. A *National Geographic* cover article from November 2005, entitled *The Secrets of Living Longer*, also recommends a whole plant food diet. The article is a lifestyle survey of three populations, Sardinians, Okinawans, and Adventists, who generally display longevity and "suffer a fraction of the diseases that commonly kill people in other parts of the developed world, and enjoy more healthy years of life." In common with all three groups is to "Eat fruits, vegetables, and whole grains."

The *National Geographic* article noted that an NIH funded study of 34,000 Seventh-day Adventists between 1976 and 1988 "...found that the Adventists' habit of consuming beans, soy milk, tomatoes, and other fruits lowered their risk of developing certain cancers. It also suggested that eating whole grain bread, drinking five glasses of water a day, and, most surprisingly, consuming four servings of nuts a week reduced their risk of heart disease."

The French "Paradox"

The French paradox is the observation that the French suffer a relatively low incidence of coronary heart disease, despite having a diet relatively rich in saturated fats. A number of explanations have been suggested:

- Saturated fat consumption does not cause heart disease
- Reduced consumption of processed carbohydrate and other junk foods.
- Regular consumption of red wine.
- Higher consumption of artificially produced trans-fats by Americans, which has been shown to have greater lipoprotein effects per gram than saturated fat.

However, statistics collected by the World Health Organization from 1990–2000 show that the incidence of heart disease in France may have been underestimated and, in fact, may be similar to that of neighboring countries.

Phytochemicals

Phytochemicals such as polyphenols are compounds produced naturally in plants (phyto means "plant" in Greek). In general, the term is used to refer to those chemicals under research to assess whether they have biological significance. To date, there is no evidence in humans that polyphenols or other non-nutrient compounds from plants have health effects. Further, dietary supplements containing phytochemicals also have no proven health benefit.

Blackberries are a source of polyphenols

While initial studies sought to reveal if antioxidant or phytochemical supplements might promote health, later large clinical trials did not detect any benefit and showed instead that excess supplementation could be harmful.

Colorful fruits and vegetables can be components of a healthy diet.

Intestinal Bacterial Flora

Animal intestines contain a large population of gut flora. In humans, the four dominant phyla are Firmicutes, Bacteroidetes, Actinobacteria, and Proteobacteria. They are essential to digestion and are also affected by food that is consumed. Bacteria in the gut perform many important functions for humans, including breaking down and aiding in the absorption of otherwise indigestible food; stimulating cell growth; repressing the growth of harmful bacteria, training the immune system to respond only to pathogens; producing vitamin B_{12}; and defending against some infectious diseases.

Animal Nutrition

Nutritional science investigates the metabolic and physiological responses of the body to diet. With advances in the fields of molecular biology, biochemistry, nutritional immunology, molecular medicine and genetics, the study of nutrition is increasingly concerned with metabolism and metabolic pathways: the sequences of biochemical steps through which substances in living things change from one form to another.

Carnivore and herbivore diets are contrasting, with basic nitrogen and carbon proportions vary for their particular foods. "The nitrogen content of plant tissues averages about 2%, while in fungi, animals, and bacteria it averages about 5% to 10%." Many herbivores rely on bacterial fermentation to create digestible nutrients from indigestible plant cellulose, while obligate carnivores must eat animal meats to obtain certain vitamins or nutrients their bodies cannot otherwise synthesize. All animals' diets must provide sufficient amounts of the basic building blocks they need, up to the point where their particular biology can synthesize the rest. Animal tissue contains chemical compounds, such as water, carbohydrates (sugar, starch, and fiber), amino acids (in proteins), fatty acids (in lipids), and nucleic acids (DNA and RNA). These compounds in turn consist of elements such as carbon, hydrogen, oxygen, nitrogen, phosphorus, calcium, iron, zinc, magnesium, manganese, and so on. All of these chemical compounds and elements occur in various forms and combinations (e.g. hormones, vitamins, phospholipids, hydroxyapatite).

Animal tissue consists of elements and compounds ingested, digested, absorbed, and circulated through the bloodstream to feed the cells of the body. Except in the unborn fetus, the digestive system is the first system involved. Digestive juices break chemical bonds in ingested molecules, and modify their conformations and energy states. Though some molecules are absorbed into the bloodstream unchanged, digestive processes release them from the matrix of foods. Unabsorbed matter, along with some waste products of metabolism, is eliminated from the body in the feces.

Studies of nutritional status must take into account the state of the body before and after experiments, as well as the chemical composition of the whole diet and of all material excreted and eliminated from the body (in urine and feces). Comparing the food to the waste can help determine the specific compounds and elements absorbed and metabolized in the body. The effects of nutrients may only be discernible over an extended period, during which all food and waste must be analyzed. The number of variables involved in such experiments is high, making nutritional studies time-consuming and expensive, which explains why the science of animal nutrition is still slowly evolving.

In particular, the consumption of whole-plant foods slows digestion and allows better absorption, and a more favorable balance of essential nutrients per Calorie, resulting in better management of cell growth, maintenance, and mitosis (cell division), as well as better regulation of appetite and blood sugar. Regularly scheduled meals (every few hours) have also proven more wholesome than infrequent or haphazard ones.

Plant Nutrition

Plant nutrition is the study of the chemical elements that are necessary for plant growth. There are several principles that apply to plant nutrition. Some elements are directly involved in plant metabolism. However, this principle does not account for the so-called beneficial elements, whose presence, while not required, has clear positive effects on plant growth.

A nutrient that is able to limit plant growth according to Liebig's law of the minimum is considered an essential plant nutrient if the plant cannot complete its full life cycle without it. There are 16 essential plant soil nutrients, besides the three major elemental nutrients carbon and oxygen that are obtained by photosynthetic plants from carbon dioxide in air, and hydrogen, which is obtained from water.

Plants uptake essential elements from the soil through their roots and from the air (consisting of mainly nitrogen and oxygen) through their leaves. Green plants obtain their carbohydrate supply from the carbon dioxide in the air by the process of photosynthesis. Carbon and oxygen are absorbed from the air, while other nutrients are absorbed from the soil. Nutrient uptake in the soil is achieved by cation exchange, wherein root hairs pump hydrogen ions (H^+) into the soil through proton pumps. These hydrogen ions displace cations attached to negatively charged soil particles so that the cations are available for uptake by the root. In the leaves, stomata open to take in carbon dioxide and expel oxygen. The carbon dioxide molecules are used as the carbon source in photosynthesis.

Although nitrogen is plentiful in the Earth's atmosphere, very few plants can use this directly. Most plants, therefore, require nitrogen compounds to be present in the soil in which they grow. This is made possible by the fact that largely inert atmospheric nitrogen is changed in a nitrogen fixation process to biologically usable forms in the soil by bacteria.

Plant nutrition is a difficult subject to understand completely, partially because of the variation between different plants and even between different species or individuals of a given clone. Elements present at low levels may cause deficiency symptoms, and toxicity is possible at levels that are too high. Furthermore, deficiency of one element may present as symptoms of toxicity from another element, and vice versa.

Environmental Nutrition

Research in the field of nutrition has greatly contributed in finding out the essential facts about how environmental depletion can lead to crucial nutrition-related health problems like contamination, spread of contagious diseases, malnutrition, etc. Moreover, environmental contamination due to discharge of agricultural as well as industrial chemicals like organocholrines, heavy metal, and radionucleotides may adversely affect the human and the ecosystem as a whole. As far as safety of the human health is concerned, then these environmental contaminants can reduce people's nutritional status and health. This could directly or indirectly cause drastic changes in their diet habits. Hence, food-based remedial as well as preventive strategies are essential to address global issues like hunger and malnutrition and to enable the susceptible people to adapt themselves to all these environmental as well as socio-economic alterations.

Advice and Guidance

Government Policies

Canada's Food Guide is another example government-run nutrition program. Produced by Health Canada, the guide advises food quantities, provides education on balanced nutrition, and promotes physical activity in accordance with government-mandated nutrient needs. Like other nutrition programs around the world, Canada's Food Guide divides nutrition into four main food groups: vegetables and fruit, grain products, milk and alternatives, and meat and alternatives.It is interesting to note that, unlike its American counterpart, the Canadian guide references and provides alternative to meat and dairy, which can be attributed to the growing vegan and vegetarian movements.

In the US, dietitians are registered (RD) or licensed (LD) with the Commission for Dietetic Registration and the American Dietetic Association, and are only able to use the title "dietitian," as described by the business and professions codes of each respective state, when they have met specific educational and experiential prerequisites and passed a national registration or licensure examination, respectively. In California, registered dietitians must abide by the *"Business and Professions Code of Section 2585-2586.8"*.Anyone may call themselves a nutritionist, including unqualified dietitians, as this term is unregulated. Some states, such as the State of Florida, have begun to include the title "nutritionist" in state licensure requirements. Most governments provide guidance on nutrition, and some also impose mandatory disclosure/labeling requirements for processed food manufacturers and restaurants to assist consumers in complying with such guidance.

In the US, nutritional standards and recommendations are established jointly by the US Department of Agriculture and US Department of Health and Human Services. Dietary and physical activity guidelines from the USDA are presented in the concept of MyPlate, which superseded the food pyramid, which replaced the Four Food Groups. The Senate committee currently responsible for oversight of the USDA is the *Agriculture, Nutrition and Forestry Committee*. Committee hearings are often televised on C-SPAN. The U.S. Department of Health and Human Services provides a sample week-long menu that fulfills the nutritional recommendations of the government.

Government Programs

Governmental organisations have been working on nutrition literacy interventions in non-primary health care settings to address the nutrition information problem in the U.S. Some programs include:

The Family Nutrition Program (FNP) is a free nutrition education program serving low-income adults around the U.S. This program is funded by the Food Nutrition Service's (FNS) branch of the United States Department of Agriculture (USDA) usually through a local state academic institution that runs the program. The FNP has developed a series of tools to help families participating in the Food Stamp Program stretch their food dollar and form healthful eating habits including nutrition education.

Expanded Food and Nutrition Education Program (ENFEP) is a unique program that currently operates in all 50 states and in American Samoa, Guam, Micronesia, Northern Marianas, Puerto Rico, and the Virgin Islands. It is designed to assist limited-resource audiences in acquiring the knowledge, skills, attitudes, and changed behavior necessary for nutritionally sound diets, and to contribute to their personal development and the improvement of the total family diet and nutritional well-being.

An example of a state initiative to promote nutrition literacy is Smart Bodies, a public-private partnership between the state's largest university system and largest health insurer, Louisiana State Agricultural Center and Blue Cross and Blue Shield of Louisiana Foundation. Launched in 2005, this program promotes lifelong healthful eating patterns and physically active lifestyles for children and their families. It is an interactive educational program designed to help prevent childhood obesity through classroom activities that teach children healthful eating habits and physical exercise.

Education

Nutrition is taught in schools in many countries. In England and Wales, the Personal and Social Education and Food Technology curricula include nutrition, stressing the importance of a balanced diet and teaching how to read nutrition labels on packaging. In many schools, a Nutrition class will fall within the Family and Consumer Science or Health departments. In some American schools, students are required to take a certain number of FCS or Health related classes. Nutrition is offered at many schools, and, if it is not a class of its own, nutrition is included in other FCS or Health classes such as: Life Skills, Independent Living, Single Survival, Freshmen Connection, Health etc. In many Nutrition classes, students learn about the food groups, the food pyramid, Daily Recommended Allowances, calories, vitamins, minerals, malnutrition, physical activity, healthful food choices, portion sizes, and how to live a healthy life.

A 1985, US National Research Council report entitled *Nutrition Education in US Medical Schools* concluded that nutrition education in medical schools was inadequate.

Only 20% of the schools surveyed taught nutrition as a separate, required course. A 2006 survey found that this number had risen to 30%.

Nutrition Literacy

At the time of this entry, we were not able to identify any specific nutrition literacy studies in the U.S. at a national level. However, the findings of the 2003 National Assessment of Adult Literacy (NAAL) provide a basis upon which to frame the nutrition literacy problem in the U.S. NAAL introduced the first ever measure of "the degree to which individuals have the capacity to obtain, process and understand basic health information and services needed to make appropriate health decisions" – an objective of Healthy People 2010 and of which nutrition literacy might be considered an important subset. On a scale of below basic, basic, intermediate and proficient, NAAL found 13 percent of adult Americans have proficient health literacy, 44% have intermediate literacy, 29 percent have basic literacy and 14 percent have below basic health literacy. The study found that health literacy increases with education and people living below the level of poverty have lower health literacy than those above it.

Another study examining the health and nutrition literacy status of residents of the lower Mississippi Delta found that 52 percent of participants had a high likelihood of limited literacy skills. While a precise comparison between the NAAL and Delta studies is difficult, primarily because of methodological differences, Zoellner et al. suggest that health literacy rates in the Mississippi Delta region are different from the U.S. general population and that they help establish the scope of the problem of health literacy among adults in the Delta region. For example, only 12 percent of study participants identified the My Pyramid graphic two years after it had been launched by the USDA. The study also found significant relationships between nutrition literacy and income level and nutrition literacy and educational attainment further delineating priorities for the region.

These statistics point to the complexities surrounding the lack of health/nutrition literacy and reveal the degree to which they are embedded in the social structure and interconnected with other problems. Among these problems are the lack of information about food choices, a lack of understanding of nutritional information and its application to individual circumstances, limited or difficult access to healthful foods, and a range of cultural influences and socioeconomic constraints such as low levels of education and high levels of poverty that decrease opportunities for healthful eating and living.

The links between low health literacy and poor health outcomes has been widely documented and there is evidence that some interventions to improve health literacy have produced successful results in the primary care setting. More must be done to further our understanding of nutrition literacy specific interventions in non-primary care settings in order to achieve better health outcomes.

Malnutrition

Malnutrition refers to insufficient, excessive, or imbalanced consumption of nutrients by an organism. In developed countries, the diseases of malnutrition are most often associated with nutritional imbalances or excessive consumption. In developing countries, malnutrition is more likely to be caused by poor access to a range of nutritious foods or inadequate knowledge. In Mali the International Crops Research Institute for the Semi-Arid Tropics (ICRISAT) and the Aga Khan Foundation, trained women's groups to make *equinut*, a healthy and nutritional version of the traditional recipe *di-dègué* (comprising peanut paste, honey and millet or rice flour). The aim was to boost nutrition and livelihoods by producing a product that women could make and sell, and which would be accepted by the local community because of its local heritage.

Although there are more organisms in the world who are malnourished due to insufficient consumption, increasingly more organisms suffer from excessive over-nutrition; a problem caused by an over abundance of sustenance coupled with the instinctual desire (by animals in particular) to consume all that it can.

Nutritionism is the view that excessive reliance on food science and the study of nutrition can lead to poor nutrition and to ill health. It was originally credited to Gyorgy Scrinis, and was popularized by Michael Pollan. Since nutrients are invisible, policy makers rely on nutrition experts to advise on food choices. Because science has an incomplete understanding of how food affects the human body, Pollan argues, nutritionism can be blamed for many of the health problems relating to diet in the Western World today.

Insufficient

In general, *under-consumption* refers to the long-term consumption of insufficient sustenance in relation to the energy that an organism expends or expels, leading to poor health.

Excessive

In general, *over-consumption* refers to the long-term consumption of excess sustenance in relation to the energy that an organism expends or expels, leading to poor health and, in animals, obesity. It can cause excessive hair loss, brittle nails, and irregular premenstrual cycles for females.

Unbalanced

When too much of one or more nutrients is present in the diet to the exclusion of the proper amount of other nutrients, the diet is said to be unbalanced.

Illnesses Caused by Improper Nutrient Consumption

Nutrients	Deficiency	Excess
Macronutrients		
Calories	Starvation, marasmus	Obesity, diabetes mellitus, cardiovascular disease
Simple carbohydrates	Low energy levels.	Obesity, diabetes mellitus, cardiovascular disease
Complex carbohydrates	Micronutrient deficiency	Obesity, cardiovascular disease (high glycemic index foods)
Protein	Kwashiorkor	Rabbit starvation, ketoacidosis (in diabetics)
Saturated fat	Low testosterone levels, vitamin deficiencies.	Obesity, cardiovascular disease
Trans fat	None	Obesity, cardiovascular disease
Unsaturated fat	Fat-soluble vitamin deficiency	Obesity, cardiovascular disease
Micronutrients		
Vitamin A	Xerophthalmia and night blindness	Hypervitaminosis A (cirrhosis, hair loss)
Vitamin B_1	Beri-Beri	?
Vitamin B_2	Skin and corneal lesions	?
Niacin	Pellagra	Dyspepsia, cardiac arrhythmias, birth defects
Vitamin B_{12}	Pernicious anemia	?
Vitamin C	Scurvy	Diarrhea causing dehydration
Vitamin D	Rickets	Hypervitaminosis D (dehydration, vomiting, constipation)
Vitamin E	Neurological disease	Hypervitaminosis E (anticoagulant: excessive bleeding)
Vitamin K	Hemorrhage	Liver damage
Omega-3 fats	Cardiovascular Disease	Bleeding, Hemorrhages, Hemorrhagic stroke, reduced glycemic control among diabetics
Omega-6 fats	None	Cardiovascular Disease, Cancer
Cholesterol	None	Cardiovascular Disease
Macrominerals		
Calcium	Osteoporosis, tetany, carpopedal spasm, laryngospasm, cardiac arrhythmias	Fatigue, depression, confusion, nausea, vomiting, constipation, pancreatitis, increased urination, kidney stones
Magnesium	Hypertension	Weakness, nausea, vomiting, impaired breathing, and hypotension
Potassium	Hypokalemia, cardiac arrhythmias	Hyperkalemia, palpitations
Sodium	Hyponatremia	Hypernatremia, hypertension

Trace minerals		
Iron	Anemia	Cirrhosis, Hereditary hemochromatosis, heart disease
Iodine	Goiter, hypothyroidism	Iodine toxicity (goiter, hypothyroidism)

Mental Agility

Research indicates that improving the awareness of nutritious meal choices and establishing long-term habits of healthy eating have a positive effect on cognitive and spatial memory capacity, with potential to increase a student's ability to process and retain academic information.

Some organizations have begun working with teachers, policymakers, and managed foodservice contractors to mandate improved nutritional content and increased nutritional resources in school cafeterias from primary to university level institutions. Health and nutrition have been proven to have close links with overall educational success. Currently, less than 10% of American college students report that they eat the recommended five servings of fruit and vegetables daily. Better nutrition has been shown to affect both cognitive and spatial memory performance; a study showed those with higher blood sugar levels performed better on certain memory tests. In another study, those who consumed yogurt performed better on thinking tasks when compared to those that consumed caffeine-free diet soda or confections. Nutritional deficiencies have been shown to have a negative effect on learning behavior in mice as far back as 1951.

> "Better learning performance is associated with diet-induced effects on learning and memory ability".
>
> The "nutrition-learning nexus" demonstrates the correlation between diet and learning and has application in a higher education setting.
>
> "We find that better-nourished children perform significantly better in school, partly because they enter school earlier and thus have more time to learn but mostly because of greater learning productivity per year of schooling."
>
> 91% of college students feel that they are in good health, whereas only 7% eat their recommended daily allowance of fruits and vegetables.
>
> Nutritional education is an effective and workable model in a higher education setting.
>
> More "engaged" learning models that encompass nutrition is an idea that is picking up steam at all levels of the learning cycle.

There is limited research available that directly links a student's Grade Point Average (G.P.A.) to their overall nutritional health. Additional substantive data is needed to prove that overall intellectual health is closely linked to a person's diet, rather than just another correlation fallacy.

Mental Disorders

Nutritional supplement treatment may be appropriate for major depression, bipolar disorder, schizophrenia, and obsessive compulsive disorder, the four most common mental disorders in developed countries. Supplements that have been studied most for mood elevation and stabilization include eicosapentaenoic acid and docosahexaenoic acid (each of which an omega-3 fatty acid contained in fish oil but not in flaxseed oil), vitamin B12, folic acid, and inositol.

Cancer

Cancer is now common in developing countries. According to a study by the International Agency for Research on Cancer, "In the developing world, cancers of the liver, stomach and esophagus were more common, often linked to consumption of carcinogenic preserved foods, such as smoked or salted food, and parasitic infections that attack organs." Lung cancer rates are rising rapidly in poorer nations because of increased use of tobacco. Developed countries "tended to have cancers linked to affluence or a 'Western lifestyle' — cancers of the colon, rectum, breast and prostate — that can be caused by obesity, lack of exercise, diet and age."

Metabolic Syndrome

Several lines of evidence indicate lifestyle-induced hyperinsulinemia and reduced insulin function (i.e., insulin resistance) as a decisive factor in many disease states. For example, hyperinsulinemia and insulin resistance are strongly linked to chronic inflammation, which in turn is strongly linked to a variety of adverse developments such as arterial microinjuries and clot formation (i.e., heart disease) and exaggerated cell division (i.e., cancer). Hyperinsulinemia and insulin resistance (the so-called metabolic syndrome) are characterized by a combination of abdominal obesity, elevated blood sugar, elevated blood pressure, elevated blood triglycerides, and reduced HDL cholesterol. The negative effect of hyperinsulinemia on prostaglandin PGE1/PGE2 balance may be significant.

The state of obesity clearly contributes to insulin resistance, which in turn can cause type 2 diabetes. Virtually all obese and most type 2 diabetic individuals have marked insulin resistance. Although the association between overweight and insulin resistance is clear, the exact (likely multifarious) causes of insulin resistance remain less clear. It is important to note that it has been demonstrated that appropriate exercise, more regular food intake, and reducing glycemic load all can reverse insulin re-sistance in overweight individuals (and thereby lower blood sugar levels in those with type 2 diabetes).

Obesity can unfavourably alter hormonal and metabolic status via resistance to the hormone leptin, and a vicious cycle may occur in which insulin/leptin resistance and obesity aggravate one another. The vicious cycle is putatively fuelled by continuously high insulin/leptin stimulation and fat storage, as a result of high intake of strongly insulin/leptin stimulating foods and energy. Both insulin and leptin normally function as satiety signals to the hypothalamus in the brain; however, insulin/leptin resistance may reduce this signal and therefore allow continued overfeeding despite large body fat stores. In addition, reduced leptin signalling to the brain may reduce leptin's normal effect to maintain an appropriately high metabolic rate.

There is a debate about how and to what extent different dietary factors— such as intake of processed carbohydrates, total protein, fat, and carbohydrate intake, intake of saturated and trans fatty acids, and low intake of vitamins/minerals—contribute to the development of insulin and leptin resistance. In any case, analogous to the way modern man-made pollution may possess the potential to overwhelm the environment's ability to maintain homeostasis, the recent explosive introduction of high glycemic index and processed foods into the human diet may possess the potential to overwhelm the body's ability to maintain homeostasis and health (as evidenced by the metabolic syndrome epidemic).

Hyponatremia

Excess water intake, without replenishment of sodium and potassium salts, leads to hyponatremia, which can further lead to water intoxication at more dangerous levels. A well-publicized case occurred in 2007, when Jennifer Strange died while participating in a water-drinking contest. More usually, the condition occurs in long-distance endurance events (such as marathon or triathlon competition and training) and causes gradual mental dulling, headache, drowsiness, weakness, and confusion; extreme cases may result in coma, convulsions, and death. The primary damage comes from swelling of the brain, caused by increased osmosis as blood salinity decreases. Effective fluid replacement techniques include water aid stations during running/cycling races, trainers providing water during team games, such as soccer, and devices such as Camel Baks, which can provide water for a person without making it too hard to drink the water.

Antinutrient

Antinutrients are natural or synthetic compounds that interfere with the absorption of nutrients. Nutrition studies focus on antinutrients commonly found in food sources and beverages.

Sugar Consumption in the United States

The relatively recent increased consumption of sugar has been linked to the rise of some afflictions such as diabetes, obesity, and more recently heart disease. Increased consumption of sugar has been tied to these three, among others. Obesity levels have more than doubled in the last 30 years among adults, going from 15% to 35% in the United States. Obesity and diet also happen to be high risk factors for diabetes. In the same time span that obesity doubled, diabetes numbers quadrupled in America. Increased weight, especially in the form of belly fat, and high sugar intake are also high risk factors for heart disease. Both sugar intake and fatty tissue increase the probability of elevated LDL cholesterol in the bloodstream. Elevated amounts of Low-density lipoprotein (LDL) cholesterol, is the primary factor in heart disease. In order to avoid all the dangers of sugar, moderate consumption is paramount.

Processed Foods

Since the Industrial Revolution some two hundred years ago, the food processing industry has invented many technologies that both help keep foods fresh longer and alter the fresh state of food as they appear in nature. Cooling is the primary technology used to maintain freshness, whereas many more technologies have been invented to allow foods to last longer without becoming spoiled. These latter technologies include pasteurisation, autoclavation, drying, salting, and separation of various components, all of which appearing to alter the original nutritional contents of food. Pasteurisation and autoclavation (heating techniques) have no doubt improved the safety of many common foods, preventing epidemics of bacterial infection. But some of the (new) food processing technologies have downfalls as well.

Modern separation techniques such as milling, centrifugation, and pressing have enabled concentration of particular components of food, yielding flour, oils, juices, and so on, and even separate fatty acids, amino acids, vitamins, and minerals. Inevitably, such large-scale concentration changes the nutritional content of food, saving certain nutrients while removing others. Heating techniques may also reduce food's content of many heat-labile nutrients such as certain vitamins and phytochemicals, and possibly other yet-to-be-discovered substances. Because of reduced nutritional value, processed foods are often 'enriched' or 'fortified' with some of the most critical nutrients (usually certain vitamins) that were lost during processing. Nonetheless, processed foods tend to have an inferior nutritional profile compared to whole, fresh foods, regarding content of both sugar and high GI starches, potassium/sodium, vitamins, fiber, and of intact, unoxidized (essential) fatty acids. In addition, processed foods often contain potentially harmful substances such as oxidized fats and trans fatty acids.

A dramatic example of the effect of food processing on a population's health is the history of epidemics of beri-beri in people subsisting on polished rice. Removing the outer layer of rice by polishing it removes with it the essential vitamin thiamine, causing beri-beri. Another example is the development of scurvy among infants in the late 19th century in the United States. It turned out that the vast majority of sufferers were being fed milk that had been heat-treated (as suggested by Pasteur) to control bacterial disease. Pasteurisation was effective against bacteria, but it destroyed the vitamin C.

As mentioned, lifestyle- and obesity-related diseases are becoming increasingly prevalent all around the world. There is little doubt that the increasingly widespread application of some modern food processing technologies has contributed to this development. The food processing industry is a major part of modern economy, and as such it is influential in political decisions (e.g., nutritional recommendations, agricultural subsidising). In any known profit-driven economy, health considerations are hardly a priority; effective production of cheap foods with a long shelf-life is more the trend. In general, whole, fresh foods have a relatively short shelf-life and are less profitable to produce and sell than are more processed foods. Thus, the consumer is left with the choice between more expensive, but nutritionally superior, whole, fresh foods, and cheap, usually nutritionally inferior, processed foods. Because processed foods are often cheaper, more convenient (in both purchasing, storage, and preparation), and more available, the consumption of nutritionally inferior foods has been increasing throughout the world along with many nutrition-related health complications.

Biodiversity

Biodiversity, a contraction of "biological diversity," generally refers to the variety and variability of life on Earth. One of the most widely used definitions defines it in terms of the variability within species, between species and between ecosystems. It is a measure of the variety of organisms present in different ecosystems. This can refer to genetic variation, ecosystem variation, or species variation (number of species) within an area, biome, or planet. Terrestrial biodiversity tends to be greater near the equator, which seems to be the result of the warm climate and high primary productivity. Biodiversity is not distributed evenly on Earth. It is richest in the tropics. Marine biodiversity tends to be highest along coasts in the Western Pacific, where sea surface temperature is highest and in the mid-latitudinal band in all oceans. There are latitudinal gradients in species diversity. Biodiversity generally tends to cluster in hotspots, and has been increasing through time, but will be likely to slow in the future.

The number and variety of plants, animals and other organisms that exist is known as biodiversity. It is an essential component of nature and it ensures the survival of human species by providing food, fuel, shelter, medicines and other resources to mankind. The richness of biodiversity depends on the climatic conditions and area of the region. All species of plants taken together are known as flora and about 70,000 species of plants are known to date. All species of animals taken together are known as fauna which includes birds, mammals, fish, reptiles, insects, crustaceans, molluscs, etc.

Rapid environmental changes typically cause mass extinctions. More than 99 percent of all species, amounting to over five billion species, that ever lived on Earth are estimated to be extinct. Estimates on the number of Earth's current species range from 10 million to 14 million, of which about 1.2 million have been documented and over 86 percent have not yet been described. More recently, in May 2016, scientists reported that 1 trillion species are estimated to be on Earth currently with only one-thousandth of one percent described. The total amount of related DNA base pairs on Earth is estimated at 5.0×10^{37} and weighs 50 billion tonnes. In comparison, the total mass of the biosphere has been estimated to be as much as 4 TtC (trillion tons of carbon). In July 2016, scientists reported identifying a set of 355 genes from the Last Universal Common Ancestor (LUCA) of all organisms living on Earth.

The age of the Earth is about 4.54 billion years old. The earliest undisputed evidence of life on Earth dates at least from 3.5 billion years ago, during the Eoarchean Era after a geological crust started to solidify following the earlier molten Hadean Eon. There are microbial mat fossils found in 3.48 billion-year-old sandstone discovered in Western Australia. Other early physical evidence of a biogenic substance is graphite in 3.7 billion-year-old meta-sedimentary rocks discovered in Western Greenland. More recently, in 2015, "remains of biotic life" were found in 4.1 billion-year-old rocks in Western Australia. According to one of the researchers, "If life arose relatively quickly on Earth .. then it could be common in the universe."

Since life began on Earth, five major mass extinctions and several minor events have led to large and sudden drops in biodiversity. The Phanerozoic eon (the last 540 million years) marked a rapid growth in biodiversity via the Cambrian explosion—a period during which the majority of multicellular phyla first appeared. The next 400 million years included repeated, massive biodiversity losses classified as mass extinction events. In the Carboniferous, rainforest collapse led to a great loss of plant and animal life. The Permian–Triassic extinction event, 251 million years ago, was the worst; vertebrate recovery took 30 million years. The most recent, the Cretaceous–Paleogene extinction event, occurred 65 million years ago and has often attracted more attention than others because it resulted in the extinction of the dinosaurs.

The period since the emergence of humans has displayed an ongoing biodiversity reduction and an accompanying loss of genetic diversity. Named the Holocene extinction, the reduction is caused primarily by human impacts, particularly habitat destruction. Conversely, biodiversity impacts human health in a number of ways, both positively and negatively.

The United Nations designated 2011–2020 as the United Nations Decade on Biodiversity.

Etymology

The term biological diversity was used first by wildlife scientist and conservationist Raymond F. Dasmann in the year 1968 lay book *A Different Kind of Country* advocating conservation. The term was widely adopted only after more than a decade, when in the 1980s it came into common usage in science and environmental policy. Thomas Lovejoy, in the foreword to the book *Conservation Biology*, introduced the term to the scientific community. Until then the term "natural diversity" was common, introduced by The Science Division of The Nature Conservancy in an important 1975 study, "The Preservation of Natural Diversity." By the early 1980s TNC's Science program and its head, Robert E. Jenkins, Lovejoy and other leading conservation scientists at the time in America advocated the use of the term "biological diversity".

The term's contracted form *biodiversity* may have been coined by W.G. Rosen in 1985 while planning the 1986 *National Forum on Biological Diversity* organized by the National Research Council (NRC). It first appeared in a publication in 1988 when sociobiologist E. O. Wilson used it as the title of the proceedings of that forum.

Since this period the term has achieved widespread use among biologists, environmentalists, political leaders and concerned citizens.

A similar term in the United States is *"natural heritage."* It pre-dates the others and is more accepted by the wider audience interested in conservation. Broader than biodiversity, it includes geology and landforms.

Definitions

A sampling of fungi collected during summer 2008 in Northern Saskatchewan mixed woods, near LaRonge is an example regarding the species diversity of fungi. In this photo, there are also leaf lichens and mosses.

"Biodiversity" is most commonly used to replace the more clearly defined and long established terms, species diversity and species richness. Biologists most often define biodiversity as the "totality of genes, species and ecosystems of a region". An advantage of this definition is that it seems to describe most circumstances and presents a unified view of the traditional types of biological variety previously identified:

- taxonomic diversity (usually measured at the species diversity level)
- ecological diversity often viewed from the perspective of ecosystem diversity
- morphological diversity which stems from genetic diversity
- functional diversity which is a measure of the number of functionally disparate species within a population (e.g. different feeding mechanism, different motility, predator vs prey, etc.)

In 2003 Professor Anthony Campbell at Cardiff University, UK and the Darwin Centre, Pembrokeshire, defined a fourth level: Molecular Diversity.

This multilevel construct is consistent with Datman and Lovejoy. An explicit definition consistent with this interpretation was first given in a paper by Bruce A. Wilcox commissioned by the International Union for the Conservation of Nature and Natural Resources (IUCN) for the 1982 World National Parks Conference. Wilcox's definition was "Biological diversity is the variety of life forms...at all levels of biological systems (i.e., molecular, organismic, population, species and ecosystem)...". The 1992 United Nations Earth Summit defined "biological diversity" as "the variability among living organisms from all sources, including, 'inter alia', terrestrial, marine and other aquatic ecosystems and the ecological complexes of which they are part: this includes diversity within species, between species and of ecosystems". This definition is used in the United Nations Convention on Biological Diversity.

One textbook's definition is "variation of life at all levels of biological organization".

Genetically biodiversity can be defined as the diversity of alleles, genes and organisms. They study processes such as mutation and gene transfer that drive evolution.

Measuring diversity at one level in a group of organisms may not precisely correspond to diversity at other levels. However, tetrapod (terrestrial vertebrates) taxonomic and ecological diversity shows a very close correlation.

Distribution

Biodiversity is not evenly distributed, rather it varies greatly across the globe as well as within regions. Among other factors, the diversity of all living things (biota) depends on temperature, precipitation, altitude, soils, geography and the presence of other species. The study of the spatial distribution of organisms, species and ecosystems, is the science of biogeography.

Diversity consistently measures higher in the tropics and in other localized regions such as the Cape Floristic Region and lower in polar regions generally. Rain forests that have had wet climates for a long time, such as Yasuní National Park in Ecuador, have particularly high biodiversity.

A conifer forest in the Swiss Alps (National Park)

Terrestrial biodiversity is thought to be up to 25 times greater than ocean biodiversity. A recently discovered method put the total number of species on Earth at 8.7 million, of which 2.1 million were estimated to live in the ocean. However, this estimate seems to under-represent the diversity of microorganisms.

Latitudinal Gradients

Generally, there is an increase in biodiversity from the poles to the tropics. Thus localities at lower latitudes have more species than localities at higher latitudes. This is often referred to as the latitudinal gradient in species diversity. Several ecological mechanisms may contribute to the gradient, but the ultimate factor behind many of them is the greater mean temperature at the equator compared to that of the poles.

Even though terrestrial biodiversity declines from the equator to the poles, some studies claim that this characteristic is unverified in aquatic ecosystems, especially in marine ecosystems. The latitudinal distribution of parasites does not appear to follow this rule.

Hotspots

A biodiversity hotspot is a region with a high level of endemic species that has experienced great habitat loss. The term hotspot was introduced in 1988 by Norman Myers. While hotspots are spread all over the world, the majority are forest areas and most are located in the tropics.

Brazil's Atlantic Forest is considered one such hotspot, containing roughly 20,000 plant species, 1,350 vertebrates and millions of insects, about half of which occur nowhere else. The island of Madagascar and India are also particularly notable. Colombia is characterized by high biodiversity, with the highest rate of species by area unit worldwide and it has the largest number of endemics (species that are not found naturally anywhere else) of any country. About 10% of the species of the Earth can be found in Colombia, including over 1,900 species of bird, more than in Europe and North America combined, Colombia has 10% of the world's mammals species, 14% of the amphibian species and 18% of the bird species of the world. Madagascar dry deciduous forests and lowland rainforests possess a high ratio of endemism. Since the island separated from mainland Africa 66 million years ago, many species and ecosystems have evolved independently. Indonesia's 17,000 islands cover 735,355 square miles (1,904,560 km^2) and contain 10% of the world's flowering plants, 12% of mammals and 17% of reptiles, amphibians and birds—along with nearly 240 million people. Many regions of high biodiversity and/or endemism arise from specialized habitats which require unusual adaptations, for example, alpine environments in high mountains, or Northern European peat bogs.

Accurately measuring differences in biodiversity can be difficult. Selection bias amongst researchers may contribute to biased empirical research for modern estimates of biodiversity. In 1768, Rev. Gilbert White succinctly observed of his Selborne, Hampshire *"all nature is so full, that that district produces the most variety which is the most examined."*

Evolution and History

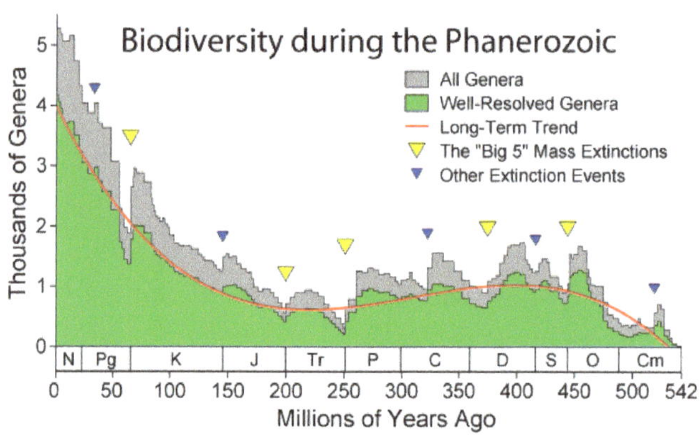

Apparent marine fossil diversity during the Phanerozoic

Biodiversity is the result of 3.5 billion years of evolution. The origin of life has not been definitely established by science, however some evidence suggests that life may already have been well-established only a few hundred million years after the formation of the Earth. Until approximately 600 million years ago, all life consisted of archaea, bacteria, protozoans and similar single-celled organisms.

The history of biodiversity during the Phanerozoic (the last 540 million years), starts with rapid growth during the Cambrian explosion—a period during which nearly every phylum of multicellular organisms first appeared. Over the next 400 million years or so, invertebrate diversity showed little overall trend and vertebrate diversity shows an overall exponential trend. This dramatic rise in diversity was marked by periodic, massive losses of diversity classified as mass extinction events. A significant loss occurred when rainforests collapsed in the carboniferous. The worst was the Permian-Triassic extinction event, 251 million years ago. Vertebrates took 30 million years to recover from this event.

The fossil record suggests that the last few million years featured the greatest biodiversity in history. However, not all scientists support this view, since there is uncertainty as to how strongly the fossil record is biased by the greater availability and preservation of recent geologic sections. Some scientists believe that corrected for sampling artifacts, modern biodiversity may not be much different from biodiversity 300 million years ago., whereas others consider the fossil record reasonably reflective of the diversification of life. Estimates of the present global macroscopic species diversity vary from 2 million to 100 million, with a best estimate of somewhere near 9 million, the vast majority arthropods. Diversity appears to increase continually in the absence of natural selection.

Evolutionary Diversification

The existence of a "global carrying capacity", limiting the amount of life that can live at once, is debated, as is the question of whether such a limit would also cap the number of species. While records of life in the sea shows a logistic pattern of growth, life on land (insects, plants and tetrapods)shows an exponential rise in diversity. As one author states, "Tetrapods have not yet invaded 64 per cent of potentially habitable modes and it could be that without human influence the ecological and taxonomic diversity of tetrapods would continue to increase in an exponential fashion until most or all of the available ecospace is filled."

On the other hand, changes through the Phanerozoic correlate much better with the hyperbolic model (widely used in population biology, demography and macrosociology, as well as fossil biodiversity) than with exponential and logistic models. The latter models imply that changes in diversity are guided by a first-order positive feedback (more ancestors, more descendants) and/or a negative feedback arising from resource limitation. Hyperbolic model implies a second-order positive feedback. The hyperbolic pattern of the world population growth arises from a second-order positive feedback between the population size and the rate of technological growth. The hyperbolic character of biodiversity growth can be similarly accounted for by a feedback between diversity and community structure complexity. The similarity between the curves of biodiversity and human population probably comes from the fact that both are derived from the interference of the hyperbolic trend with cyclical and stochastic dynamics.

Most biologists agree however that the period since human emergence is part of a new mass extinction, named the Holocene extinction event, caused primarily by the impact humans are having on the environment. It has been argued that the present rate of extinction is sufficient to eliminate most species on the planet Earth within 100 years.

New species are regularly discovered (on average between 5–10,000 new species each year, most of them insects) and many, though discovered, are not yet classified (estimates are that nearly 90% of all arthropods are not yet classified). Most of the terrestrial diversity is found in tropical forests and in general, land has more species than the ocean; some 8.7 million species may exists on Earth, of which some 2.1 million live in the ocean

Biodiversity and Ecosystem Services

Summer field in Belgium (Hamois). The blue flowers are *Centaurea cyanus* and the red are *Papaver rhoeas*.

The Balance of Evidence

"Ecosystem services are the suite of benefits that ecosystems provide to humanity."

These services come in three flavors:

1. Provisioning services which involve the production of renewable resources (e.g.: food, wood, fresh water)
2. Regulating services which are those that lessen environmental change (e.g.: climate regulation, pest/disease control)
3. Cultural services represent human value and enjoyment (e.g.: landscape aesthetics, cultural heritage, outdoor recreation and spiritual significance)

There have been many claims about biodiversity's effect on these ecosystem services, especially provisioning and regulating services. After an exhaustive survey through peer-reviewed literature to evaluate 36 different claims about biodiversity's effect on ecosystem services, 14 of those claims have been validated, 6 demonstrate mixed support or are unsupported, 3 are incorrect and 13 lack enough evidence to draw definitive conclusions.

Services Enhanced by Biodiversity

Provisioning Services

- Greater species diversity of plants increases fodder yield (synthesis of 271 experimental studies).
- Greater genetic diversity of plants (i.e.: diversity within a single species) increases overall crop yield (synthesis of 575 experimental studies). Although another review of 100 experimental studies reports mixed evidence.
- Greater species diversity of trees increases overall wood production (Synthesis of 53 experimental studies). However, there is not enough data to draw a conclusion about the effect of tree trait diversity on wood production.

Regulating Services

- Greater species diversity of fish increases the stability of fisheries yield (Synthesis of 8 observational studies)
- Greater species diversity of natural pest enemies decreases herbivorous pest populations (Data from two separate reviews; Synthesis of 266 experimental and observational studies; Synthesis of 18 observational studies. Although another review of 38 experimental studies found mixed support for this claim, suggesting that in cases where mutual intraguild predation occurs, a single predatory species is often more effective
- Greater species diversity of plants decreases disease prevalence on plants (Synthesis of 107 experimental studies)
- Greater species diversity of plants increases resistance to plant invasion (Data from two separate reviews; Synthesis of 105 experimental studies; Synthesis of 15 experimental studies)
- Greater species diversity of plants increases carbon sequestration, but note that this finding only relates to actual uptake of carbon dioxide and not long term storage, Synthesis of 479 experimental studies.
- Greater species diversity of plants increases soil nutrient remineralization (Synthesis of 103 experimental studies)

- Greater species diversity of plants increases soil organic matter (Synthesis of 85 experimental studies)

Services with Mixed Evidence

Provisioning Services

- None to date

Regulating Services

- Greater species diversity of plants may or may not decrease herbivorous pest populations. Data from two separate reviews suggest that greater diversity decreases pest populations (Synthesis of 40 observational studies; Synthesis of 100 experimental studies). One review found mixed evidence (Synthesis of 287 experimental studies), while another found contrary evidence (Synthesis of 100 experimental studies)

- Greater species diversity of animals may or may not decrease disease prevalence on those animals (Synthesis of 45 experimental and observational studies), although a 2013 study offers more support showing that biodiversity may in fact enhance disease resistance within animal communities, at least in amphibian frog ponds. Many more studies must be published in support of diversity to sway the balance of evidence will be such that we can draw a general rule on this service.

- Greater species and trait diversity of plants may or may not increase long term carbon storage (Synthesis of 33 observational studies)

- Greater pollinator diversity may or may not increase pollination (Synthesis of 7 observational studies), but a publication from March 2013 suggests that increased native pollinator diversity enhances pollen deposition (although not necessarily fruit set as the authors would have you believe, for details explore their lengthy supplementary material).

Services for which Biodiversity is a Hindrance

Provisioning Services

- Greater species diversity of plants reduces primary production (Synthesis of 7 experimental studies)

Regulating Services

- Greater genetic and species diversity of a number of organisms reduces freshwater purification (Synthesis of 8 experimental studies, although an attempt by the authors to investigate the effect of detritivore diversity on freshwater purification was unsuccessful due to a lack of available evidence (only 1 observational study was found

Provisioning Services

- Effect of species diversity of plants on biofuel yield (In a survey of the literature, the investigators only found 3 studies)
- Effect of species diversity of fish on fishery yield (In a survey of the literature, the investigators only found 4 experimental studies and 1 observational study)

Regulating Services

- Effect of species diversity on the stability of biofuel yield (In a survey of the literature, the investigators did not find any studies)
- Effect of species diversity of plants on the stability of fodder yield (In a survey of the literature, the investigators only found 2 studies)
- Effect of species diversity of plants on the stability of crop yield (In a survey of the literature, the investigators only found 1 study)
- Effect of genetic diversity of plants on the stability of crop yield (In a survey of the literature, the investigators only found 2 studies)
- Effect of diversity on the stability of wood production (In a survey of the literature, the investigators could not find any studies)
- Effect of species diversity of multiple taxa on erosion control (In a survey of the literature, the investigators could not find any studies – they did however find studies on the effect of species diversity and root biomass)
- Effect of diversity on flood regulation (In a survey of the literature, the investigators could not find any studies)
- Effect of species and trait diversity of plants on soil moisture (In a survey of the literature, the investigators only found 2 studies)

Other sources have reported somewhat conflicting results and in 1997 Robert Costanza and colleagues reported the estimated global value of ecosystem services (not captured in traditional markets) at an average of $33 trillion annually.

Since the stone age, species loss has accelerated above the average basal rate, driven by human activity. Estimates of species losses are at a rate 100-10,000 times as fast as is typical in the fossil record. Biodiversity also affords many non-material benefits including spiritual and aesthetic values, knowledge systems and education.

Biodiversity and Agriculture

Amazon Rainforest in South America

Agricultural diversity can be divided into two categories: intraspecific diversity, which includes the genetic variety within a single species, like the potato (*Solanum tuberosum*) that is composed of many different forms and types (e.g.: in the U.S. we might compare russet potatoes with new potatoes or purple potatoes, all different, but all part of the same species, *S. tuberosum*).

The other category of agricultural diversity is called interspecific diversity and refers to the number and types of different species. Thinking about this diversity we might note that many small vegetable farmers grow many different crops like potatoes and also carrots, peppers, lettuce etc.

Agricultural diversity can also be divided by whether it is 'planned' diversity or 'associated' diversity. This is a functional classification that we impose and not an intrinsic feature of life or diversity. Planned diversity includes the crops which a farmer has encouraged, planted or raised (e.g.: crops, covers, symbionts and livestock, among others), which can be contrasted with the associated diversity that arrives among the crops, uninvited (e.g.: herbivores, weed species and pathogens, among others).

The control of associated biodiversity is one of the great agricultural challenges that farmers face. On monoculture farms, the approach is generally to eradicate associated diversity using a suite of biologically destructive pesticides, mechanized tools and transgenic engineering techniques, then to rotate crops. Although some polyculture farmers use the same techniques, they also employ integrated pest management strategies as well as strategies that are more labor-intensive, but generally less dependent on capital, biotechnology and energy.

Interspecific crop diversity is, in part, responsible for offering variety in what we eat. Intraspecific diversity, the variety of alleles within a single species, also offers us choice in our diets. If a crop fails in a monoculture, we rely on agricultural diversity to replant the land with something new. If a wheat crop is destroyed by a pest we may plant a hardier variety of wheat the next year, relying on intraspecific diversity. We may forgo wheat production in that area and plant a different species altogether, relying on interspecific diversity. Even an agricultural society which primarily grows monocultures, relies on biodiversity at some point.

- The Irish potato blight of 1846 was a major factor in the deaths of one million people and the emigration of about two million. It was the result of planting only two potato varieties, both vulnerable to the blight, *Phytophthora infestans*, which arrived in 1845
- When rice grassy stunt virus struck rice fields from Indonesia to India in the 1970s, 6,273 varieties were tested for resistance. Only one was resistant, an Indian variety and known to science only since 1966. This variety formed a hybrid with other varieties and is now widely grown.
- Coffee rust attacked coffee plantations in Sri Lanka, Brazil and Central America in 1970. A resistant variety was found in Ethiopia. The diseases are themselves a form of biodiversity.

Monoculture was a contributing factor to several agricultural disasters, including the European wine industry collapse in the late 19th century and the US southern corn leaf blight epidemic of 1970.

Although about 80 percent of humans' food supply comes from just 20 kinds of plants, humans use at least 40,000 species. Many people depend on these species for food, shelter and clothing. Earth's surviving biodiversity provides resources for increasing the range of food and other products suitable for human use, although the present extinction rate shrinks that potential.

Biodiversity and Human Health

The diverse forest canopy on Barro Colorado Island, Panama, yielded this display of different fruit

Biodiversity's relevance to human health is becoming an international political issue, as scientific evidence builds on the global health implications of biodiversity loss. This issue is closely linked with the issue of climate change, as many of the anticipated health risks of climate change are associated with changes in biodiversity (e.g. changes in populations and distribution of disease vectors, scarcity of fresh water, impacts on agricultural biodiversity and food resources etc.) This is because the species most likely to disappear are those that buffer against infectious disease transmission, while surviving species tend to be the ones that increase disease transmission, such as that of West Nile Virus, Lyme disease and Hantavirus, according to a study done co-authored by Felicia Keesing, an ecologist at Bard College and Drew Harvell, associate director for Environment of the Atkinson Center for a Sustainable Future (ACSF) at Cornell University.

The growing demand and lack of drinkable water on the planet presents an additional challenge to the future of human health. Partly, the problem lies in the success of water suppliers to increase supplies and failure of groups promoting preservation of water resources. While the distribution of clean water increases, in some parts of the world it remains unequal. According to *2008 World Population Data Sheet*, only 62% of least developed countries are able to access clean water.

Some of the health issues influenced by biodiversity include dietary health and nutrition security, infectious disease, medical science and medicinal resources, social and psychological health. Biodiversity is also known to have an important role in reducing disaster risk and in post-disaster relief and recovery efforts.

Biodiversity provides critical support for drug discovery and the availability of medicinal resources. A significant proportion of drugs are derived, directly or indirectly, from biological sources: at least 50% of the pharmaceutical compounds on the US market are derived from plants, animals and micro-organisms, while about 80% of the world population depends on medicines from nature (used in either modern or traditional medical practice) for primary healthcare. Only a tiny fraction of wild species has been investigated for medical potential. Biodiversity has been critical to advances throughout the field of bionics. Evidence from market analysis and biodiversity science indicates that the decline in output from the pharmaceutical sector since the mid-1980s can be attributed to a move away from natural product exploration ("bioprospecting") in favor of genomics and synthetic chemistry, indeed claims about the value of undiscovered pharmaceuticals may not provide enough incentive for companies in free markets to search for them because of the high cost of development; meanwhile, natural products have a long history of supporting significant economic and health innovation. Marine ecosystems are particularly important, although inappropriate bioprospecting can increase biodiversity loss, as well as violating the laws of the communities and states from which the resources are taken.

Biodiversity, Business and Industry

Agriculture production, pictured is a tractor and a chaser bin

Many industrial materials derive directly from biological sources. These include building materials, fibers, dyes, rubber and oil. Biodiversity is also important to the security of resources such as water, timber, paper, fiber and food. As a result, biodiversity loss is a significant risk factor in business development and a threat to long term economic sustainability.

Biodiversity, Leisure, Cultural and Aesthetic Value

Biodiversity enriches leisure activities such as hiking, birdwatching or natural history study. Biodiversity inspires musicians, painters, sculptors, writers and other artists. Many cultures view themselves as an integral part of the natural world which requires them to respect other living organisms.

Popular activities such as gardening, fishkeeping and specimen collecting strongly depend on biodiversity. The number of species involved in such pursuits is in the tens of thousands, though the majority do not enter commerce.

The relationships between the original natural areas of these often exotic animals and plants and commercial collectors, suppliers, breeders, propagators and those who promote their understanding and enjoyment are complex and poorly understood. The general public responds well to exposure to rare and unusual organisms, reflecting their inherent value.

Philosophically it could be argued that biodiversity has intrinsic aesthetic and spiritual value to mankind *in and of itself*. This idea can be used as a counterweight to the notion that tropical forests and other ecological realms are only worthy of conservation because of the services they provide.

Biodiversity and Ecological Services

Eagle Creek, Oregon hiking

Biodiversity supports many ecosystem services:

"There is now unequivocal evidence that biodiversity loss reduces the efficiency by which ecological communities capture biologically essential resources, produce biomass, decompose and recycle biologically essential nutrients... There is mounting evidence that biodiversity increases the stability of ecosystem functions through time... Diverse communities are more productive because they contain key species that have a large influence on productivity and differences in functional traits among organisms increase total resource capture... The impacts of diversity loss on ecological processes might be sufficiently large to rival the impacts of many other global drivers of environmental change... Maintaining multiple ecosystem processes at multiple places and times requires higher levels of biodiversity than does a single process at a single place and time."

It plays a part in regulating the chemistry of our atmosphere and water supply. Biodiversity is directly involved in water purification, recycling nutrients and providing fertile soils. Experiments with controlled environments have shown that humans cannot easily build ecosystems to support human needs; for example insect pollination cannot be mimicked, and that activity alone represented between $2.1-14.6 billions in 2003.

Number of Species

Species	Earth			Ocean		
	Catalogued	Predicted	±SE	Catalogued	Predicted	±SE
Eukaryotes						
Animalia	953,434	7,770,000	958,000	171,082	2,150,000	145,000
Chromista	13,033	27,500	30,500	4,859	7,400	9,640
Fungi	43,271	611,000	297,000	1,097	5,320	11,100
Plantae	215,644	298,000	8,200	8,600	16,600	9,130
Protozoa	8,118	36,400	6,690	8,118	36,400	6,690
Total	1,233,500	8,740,000	1,300,000	193,756	2,210,000	182,000
Prokaryotes						
Archaea	502	455	160	1	1	0
Bacteria	10,358	9,680	3,470	652	1,320	436
Total	10,860	10,100	3,630	653	1,320	436
Grand Total	**1,244,360**	**8,750,000**	**1,300,000**	**194,409**	**2,210,000**	**182,000**

Predictions for prokaryotes represent a lower bound because they do not consider undescribed higher taxa. For protozoa, the ocean database was substantially more complete than the database for the entire Earth so we only used the former to estimate the total number of species in this taxon. All predictions were rounded to three significant digits.
doi:10.1371/journal.pbio.1001127.t002

Discovered and predicted total number of species on land and in the oceans

According to Mora and colleagues, the total number of terrestrial species is estimated to be around 8.7 million while the number of oceanic species is much lower, estimated at 2.2 million. The authors note that these estimates are strongest for eukaryotic organisms and likely represent the lower bound of prokaryote diversity. Other estimates include:

- 220,000 vascular plants, estimated using the species-area relation method
- 0.7-1 million marine species
- 10–30 million insects; (of some 0.9 million we know today)
- 5–10 million bacteria;
- 1.5-3 million fungi, estimates based on data from the tropics, long-term non-tropical sites and molecular studies that have revealed cryptic speciation. Some 0.075 million species of fungi had been documented by 2001)
- 1 million mites
- The number of microbial species is not reliably known, but the Global Ocean Sampling Expedition dramatically increased the estimates of genetic diversity by identifying an enormous number of new genes from near-surface plankton samples at various marine locations, initially over the 2004-2006 period. The findings may eventually cause a significant change in the way science defines species and other taxonomic categories.

Since the rate of extinction has increased, many extant species may become extinct before they are described.

Species Loss Rates

No longer do we have to justify the existence of humid tropical forests on the feeble grounds that they might carry plants with drugs that cure human disease. *Gaia theory* forces us to see that they offer much more than this. Through their capacity to evapotranspirate vast volumes of water vapor, they serve to keep the planet cool by wearing a sunshade of white reflecting cloud. Their replacement by cropland could precipitate a disaster that is global in scale.

During the last century, decreases in biodiversity have been increasingly observed. In 2007, German Federal Environment Minister Sigmar Gabriel cited estimates that up to 30% of all species will be extinct by 2050. Of these, about one eighth of known plant species are threatened with extinction. Estimates reach as high as 140,000 species per year (based on Species-area theory). This figure indicates unsustainable ecological practices, because few species emerge each year. Almost all scientists acknowledge that the rate of species loss is greater now than at any time in human history, with extinctions occurring at rates hundreds of times higher than background extinction rates. As of 2012, some studies suggest that 25% of all mammal species could be extinct in 20 years.

In absolute terms, the planet has lost 52% of its biodiversity since 1970 according to a 2014 study by the World Wildlife Fund. The Living Planet Report 2014 claims that "the number of mammals, birds, reptiles, amphibians and fish across the globe is, on average, about half the size it was 40 years ago". Of that number, 39% accounts for the terrestrial wildlife gone, 39% for the marine wildlife gone and 76% for the freshwater wildlife gone. Biodiversity took the biggest hit in Latin America, plummeting 83 percent. High-income countries showed a 10% increase in biodiversity, which was canceled out by a loss in low-income countries. This is despite the fact that high-income countries use five times the ecological resources of low-income countries, which was explained as a result of process whereby wealthy nations are outsourcing resource depletion to poorer nations, which are suffering the greatest ecosystem losses.

Threats

In 2006 many species were formally classified as rare or endangered or threatened; moreover, scientists have estimated that millions more species are at risk which have not been formally recognized. About 40 percent of the 40,177 species assessed using the IUCN Red List criteria are now listed as threatened with extinction—a total of 16,119.

Jared Diamond describes an "Evil Quartet" of habitat destruction, overkill, introduced species and secondary extinctions. Edward O. Wilson prefers the acronym HIPPO, standing for Habitat destruction, Invasive species, Pollution, human over-Population and Over-harvesting. The most authoritative classification in use today is IUCN's Classification of Direct Threats which has been adopted by major international conservation organizations such as the US Nature Conservancy, the World Wildlife Fund, Conservation International and BirdLife International.

Habitat Destruction

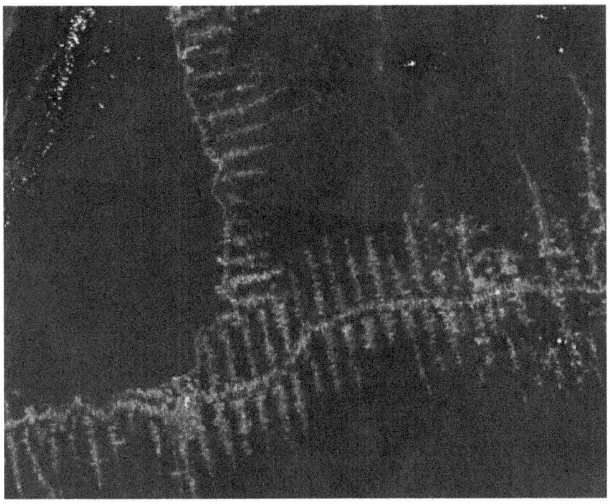

Deforestation and increased road-building in the Amazon Rainforest are a significant concern because of increased human encroachment upon wild areas, increased resource extraction and further threats to biodiversity.

Habitat destruction has played a key role in extinctions, especially related to tropical forest destruction. Factors contributing to habitat loss are: overconsumption, overpopulation, land use change, deforestation, pollution (air pollution, water pollution, soil contamination) and global warming or climate change.

Habitat size and numbers of species are systematically related. Physically larger species and those living at lower latitudes or in forests or oceans are more sensitive to reduction in habitat area. Conversion to "trivial" standardized ecosystems (e.g., monoculture following deforestation) effectively destroys habitat for the more diverse species that preceded the conversion. In some countries lack of property rights or lax law/regulatory enforcement necessarily leads to biodiversity loss (degradation costs having to be supported by the community).

A 2007 study conducted by the National Science Foundation found that biodiversity and genetic diversity are codependent—that diversity among species requires diversity within a species and vice versa. "If any one type is removed from the system, the cycle can break down and the community becomes dominated by a single species." At present, the most threatened ecosystems are found in fresh water, according to the Millennium Ecosystem Assessment 2005, which was confirmed by the "Freshwater Animal Diversity Assessment", organised by the biodiversity platform and the French *Institut de recherche pour le développement* (MNHNP).

Co-extinctions are a form of habitat destruction. Co-extinction occurs when the extinction or decline in one accompanies the other, such as in plants and beetles.

Introduced and Invasive Species

Male *Lophura nycthemera* (silver pheasant), a native of East Asia that has been introduced into parts of Europe for ornamental reasons

Barriers such as large rivers, seas, oceans, mountains and deserts encourage diversity by enabling independent evolution on either side of the barrier, via the process of allopatric speciation. The term invasive species is applied to species that breach the natural barriers that would normally keep them constrained. Without barriers, such species occupy new territory, often supplanting native species by occupying their niches, or by using resources that would normally sustain native species.

The number of species invasions has been on the rise at least since the beginning of the 1900s. Species are increasingly being moved by humans (on purpose and accidentally). In some cases the invaders are causing drastic changes and damage to their new habitats (e.g.: zebra mussels and the emerald ash borer in the Great Lakes region and the lion fish along the North American Atlantic coast). Some evidence suggests that invasive species are competitive in their new habitats because they are subject to less pathogen disturbance. Others report confounding evidence that occasionally suggest that species-rich communities harbor many native and exotic species simultaneously while some say that diverse ecosystems are more resilient and resist invasive plants and animals. An important question is, "do invasive species cause extinctions?" Many studies cite effects of invasive species on natives, but not extinctions. Invasive species seem to increase local (i.e.: alpha diversity) diversity, which decreases turnover of diversity (i.e.: beta diversity). Overall gamma diversity may be lowered because species are going extinct because of other causes, but even some of the most insidious invaders (e.g.: Dutch elm disease, emerald ash borer, chestnut blight in North America) have not caused their host species to become extinct. Extirpation, population decline and homogenization of regional biodiversity are much more common. Human activities have frequently been the cause of invasive species circumventing their barriers, by introducing them for food and other purposes. Human activities therefore allow species to migrate to new areas (and thus become invasive) occurred on time scales much shorter than historically have been required for a species to extend its range.

Not all introduced species are invasive, nor all invasive species deliberately introduced. In cases such as the zebra mussel, invasion of US waterways was unintentional. In other cases, such as mongooses in Hawaii, the introduction is deliberate but ineffective (nocturnal rats were not vulnerable to the diurnal mongoose). In other cases, such as oil palms in Indonesia and Malaysia, the introduction produces substantial economic benefits, but the benefits are accompanied by costly unintended consequences.

Finally, an introduced species may unintentionally injure a species that depends on the species it replaces. In Belgium, Prunus spinosa from Eastern Europe leafs much sooner than its West European counterparts, disrupting the feeding habits of the *Thecla betulae* butterfly (which feeds on the leaves). Introducing new species often leaves endemic and other local species unable to compete with the exotic species and unable to survive. The exotic organisms may be predators, parasites, or may simply outcompete indigenous species for nutrients, water and light.

At present, several countries have already imported so many exotic species, particularly agricultural and ornamental plants, that their own indigenous fauna/flora may be outnumbered. For example, the introduction of kudzu from Southeast Asia to Canada and the United States has threatened biodiversity in certain areas.

Genetic Pollution

Endemic species can be threatened with extinction through the process of genetic pollution, i.e. uncontrolled hybridization, introgression and genetic swamping. Genetic pollution leads to homogenization or replacement of local genomes as a result of either a numerical and/or fitness advantage of an introduced species. Hybridization and introgression are side-effects of introduction and invasion. These phenomena can be especially detrimental to rare species that come into contact with more abundant ones. The abundant species can interbreed with the rare species, swamping its gene pool. This problem is not always apparent from morphological (outward appearance) observations alone. Some degree of gene flow is normal adaptation and not all gene and genotype constellations can be preserved. However, hybridization with or without introgression may, nevertheless, threaten a rare species' existence.

Overexploitation

Overexploitation occurs when a resource is consumed at an unsustainable rate. This occurs on land in the form of overhunting, excessive logging, poor soil conservation in agriculture and the illegal wildlife trade.

About 25% of world fisheries are now overfished to the point where their current biomass is less than the level that maximizes their sustainable yield.

The overkill hypothesis, a pattern of large animal extinctions connected with human migration patterns, can be used explain why megafaunal extinctions can occur within a relatively short time period.

Hybridization, Genetic Pollution/erosion and Food Security

The Yecoro wheat (right) cultivar is sensitive to salinity, plants resulting from a hybrid cross with cultivar W4910 (left) show greater tolerance to high salinity

In agriculture and animal husbandry, the Green Revolution popularized the use of conventional hybridization to increase yield. Often hybridized breeds originated in developed countries and were further hybridized with local varieties in the developing world to create high yield strains resistant to local climate and diseases. Local governments and industry have been pushing hybridization. Formerly huge gene pools of various wild and indigenous breeds have collapsed causing widespread genetic erosion and genetic pollution. This has resulted in loss of genetic diversity and biodiversity as a whole.

(GM organisms) have genetic material altered by genetic engineering procedures such as recombinant DNA technology. GM crops have become a common source for genetic pollution, not only of wild varieties but also of domesticated varieties derived from classical hybridization.

Genetic erosion coupled with genetic pollution may be destroying unique genotypes, thereby creating a hidden crisis which could result in a severe threat to our food security. Diverse genetic material could cease to exist which would impact our ability to further hybridize food crops and livestock against more resistant diseases and climatic changes.

Climate Change

Global warming is also considered to be a major potential threat to global biodiversity in the future. For example, coral reefs - which are biodiversity hotspots - will be lost within the century if global warming continues at the current trend.

Polar bears on the sea ice of the Arctic Ocean, near the North Pole. Climate change has started affecting bear populations.

Climate change has seen many claims about potential to affect biodiversity but evidence supporting the statement is tenuous. Increasing atmospheric carbon dioxide certainly affects plant morphology and is acidifying oceans, and temperature affects species ranges, phenology, and weather, but the major impacts that have been predicted are still just *potential* impacts. We have not documented major extinctions yet, even as climate change drastically alters the biology of many species.

In 2004, an international collaborative study on four continents estimated that 10 percent of species would become extinct by 2050 because of global warming. "We need to limit climate change or we wind up with a lot of species in trouble, possibly extinct," said Dr. Lee Hannah, a co-author of the paper and chief climate change biologist at the Center for Applied Biodiversity Science at Conservation International.

A recent study predicts that up to 35% of the world terrestrial carnivores and ungulates will be at higher risk of extinction by 2050 because of the joint effects of predicted climate and land-use change under business-as-usual human development scenarios.

Human Overpopulation

From 1950 to 2011, world population increased from 2.5 billion to 7 billion and is forecast to reach a plateau of more than 9 billion during the 21st century. Sir David King, former chief scientific adviser to the UK government, told a parliamentary inquiry: *"It is self-evident that the massive growth in the human population through the 20th century has had more impact on biodiversity than any other single factor."* At least until the middle of the 21st century, worldwide losses of pristine biodiverse land will probably depend much on the worldwide human birth rate.

According to a 2014 study by the World Wildlife Fund, the global human population already exceeds planet's biocapacity - it would take the equivalent of 1.5 Earths of bio-capacity to meet our current demands. The report further points that if everyone on the planet had the Footprint of the average resident of Qatar, we would need 4.8 Earths and if we lived the lifestyle of a typical resident of the USA, we would need 3.9 Earths.

The Holocene Extinction

Rates of decline in biodiversity in this sixth mass extinction match or exceed rates of loss in the five previous mass extinction events in the fossil record. Loss of biodiversity results in the loss of natural capital that supplies ecosystem goods and services. From the perspective of the method known as Natural Economy the economic value of 17 ecosystem services for Earth's biosphere (calculated in 1997) has an estimated value of US$33 trillion ($3.3 \times 10^{13}$) per year.

Conservation

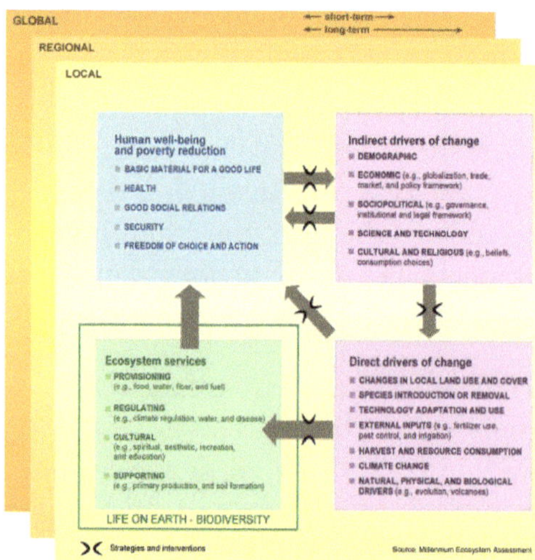

A schematic image illustrating the relationship between biodiversity, ecosystem services, human well-being and poverty. The illustration shows where conservation action, strategies and plans can influence the drivers of the current biodiversity crisis at local, regional, to global scales.

The retreat of Aletsch Glacier in the Swiss Alps (situation in 1979, 1991 and 2002), due to global warming.

Conservation biology matured in the mid-20th century as ecologists, naturalists and other scientists began to research and address issues pertaining to global biodiversity declines.

The conservation ethic advocates management of natural resources for the purpose of sustaining biodiversity in species, ecosystems, the evolutionary process and human culture and society.

Conservation biology is reforming around strategic plans to protect biodiversity. Preserving global biodiversity is a priority in strategic conservation plans that are designed to engage public policy and concerns affecting local, regional and global scales of communities, ecosystems and cultures. Action plans identify ways of sustaining human well-being, employing natural capital, market capital and ecosystem services.

In the EU Directive 1999/22/EC zoos are described as having a role in the preservation of the biodiversity of wildlife animals by conducting research or participation in breeding programs.

Protection and Restoration Techniques

Removal of exotic species will allow the species that they have negatively impacted to recover their ecological niches. Exotic species that have become pests can be identified taxonomically (e.g., with Digital Automated Identification SYstem (DAISY), using the barcode of life). Removal is practical only given large groups of individuals due to the economic cost.

As sustainable populations of the remaining native species in an area become assured, "missing" species that are candidates for reintroduction can be identified using databases such as the *Encyclopedia of Life* and the Global Biodiversity Information Facility.

- Biodiversity banking places a monetary value on biodiversity. One example is the Australian Native Vegetation Management Framework.
- Gene banks are collections of specimens and genetic material. Some banks intend to reintroduce banked species to the ecosystem (e.g., via tree nurseries).
- Reduction of and better targeting of pesticides allows more species to survive in agricultural and urbanized areas.
- Location-specific approaches may be less useful for protecting migratory species. One approach is to create wildlife corridors that correspond to the animals' movements. National and other boundaries can complicate corridor creation.

Protected Areas

Protected areas is meant for affording protection to wild animals and their habitat which also includes forest reserves and biosphere reserves. Protected areas have been set up all over the world with the specific aim of protecting and conserving plants and animals.

National Parks

National park and nature reserve is the area selected by governments or private organizations for special protection against damage or degradation with the objective of biodiversity and landscape conservation. National parks are usually owned and managed by national or state governments. A limit is placed on the number of visitors permitted to enter certain fragile areas. Designated trails or roads are created. The visitors are allowed to enter only for study, cultural and recreation purposes. Forestry operations, grazing of animals and hunting of animals are regulated. Exploitation of habitat or wildlife is banned.

Wildlife Sanctuary

Wildlife sanctuary aims only at conservation of species and have the following features:

1. The boundaries of the sanctuaries are not limited by state legislation.
2. The killing, hunting or capturing of any species is prohibited except by or under the control of the highest authority in the department which is responsible for the management of the sanctuary.
3. Private ownership may be allowed.
4. Forestry and other usages can also be permitted.

Forest Reserves

The forests play a vital role in harbouring more than 45,000 floral and 81,000 faunal species of which 5150 floral and 1837 faunal species are endemic. Plant and animal species confined to a specific geographical area are called endemic species. In reserved forests, rights to activities like hunting and grazing are sometimes given to communities living on the fringes of the forest, who sustain their livelihood partially or wholly from forest resources or products. The unclassed forests covers 6.4 percent of the total forest area and they are marked by the following characteristics:

1. They are large inaccessible forests.
2. Many of these are unoccupied.
3. They are ecologically and economically less important.

Steps to Conserve the Forest Cover

1. An extensive reforestation/afforestation program should be followed.
2. Alternative environment-friendly sources of fuel energy such as biogas other than wood should be used.
3. Loss of biodiversity due to forest fire is a major problem, immediate steps to prevent forest fire need to be taken.

4. Overgrazing by cattle can damage a forest seriously. Therefore, certain steps should be taken to prevent overgrazing by cattle.
5. Hunting and poaching should be banned.

Zoological Parks

In zoological parks or zoos, live animals are kept for public recreation, education and conservation purposes. Modern zoos offer veterinary facilities, provide opportunities for threatened species to breed in captivity and usually build environments that simulate the native habitats of the animals in their care. Zoos play a major role in creating awareness among common people about the need to conserve nature.

Botanical Gardens

Botanical garden is a garden in which plants are grown and displayed primarily for scientific and educational purposes. It consists of a collection of living plants, grown outdoors or under glass in greenhouses and conservatories. In addition, it includes a collection of dried plants or herbarium and such facilities as lecture rooms, laboratories, libraries, museums and experimental or research plantings.

Resource Allocation

Focusing on limited areas of higher potential biodiversity promises greater immediate return on investment than spreading resources evenly or focusing on areas of little diversity but greater interest in biodiversity.

A second strategy focuses on areas that retain most of their original diversity, which typically require little or no restoration. These are typically non-urbanized, non-agricultural areas. Tropical areas often fit both criteria, given their natively high diversity and relative lack of development.

Legal Status

A great deal of work is occurring to preserve the natural characteristics of Hopetoun Falls, Australia while continuing to allow visitor access.

International

- United Nations Convention on Biological Diversity (1992) and Cartagena Protocol on Biosafety;
- Convention on International Trade in Endangered Species (CITES);
- Ramsar Convention (Wetlands);
- Bonn Convention on Migratory Species;
- World Heritage Convention (indirectly by protecting biodiversity habitats)
- Regional Conventions such as the Apia Convention
- Bilateral agreements such as the Japan-Australia Migratory Bird Agreement.

Global agreements such as the Convention on Biological Diversity, give "sovereign national rights over biological resources" (not property). The agreements commit countries to "conserve biodiversity", "develop resources for sustainability" and "share the benefits" resulting from their use. Biodiverse countries that allow bioprospecting or collection of natural products, expect a share of the benefits rather than allowing the individual or institution that discovers/exploits the resource to capture them privately. Bioprospecting can become a type of biopiracy when such principles are not respected.

Sovereignty principles can rely upon what is better known as Access and Benefit Sharing Agreements (ABAs). The Convention on Biodiversity implies informed consent between the source country and the collector, to establish which resource will be used and for what and to settle on a fair agreement on benefit sharing.

National Level Laws

Biodiversity is taken into account in some political and judicial decisions:

- The relationship between law and ecosystems is very ancient and has consequences for biodiversity. It is related to private and public property rights. It can define protection for threatened ecosystems, but also some rights and duties (for example, fishing and hunting rights).
- Law regarding species is more recent. It defines species that must be protected because they may be threatened by extinction. The U.S. Endangered Species Act is an example of an attempt to address the "law and species" issue.
- Laws regarding gene pools are only about a century old. Domestication and plant breeding methods are not new, but advances in genetic engineering have led to tighter laws covering distribution of genetically modified organisms, gene patents and process patents. Governments struggle to decide whether to focus on for example, genes, genomes, or organisms and species.

Uniform approval for use of biodiversity as a legal standard has not been achieved, however. Bosselman argues that biodiversity should not be used as a legal standard, claiming that the remaining areas of scientific uncertainty cause unacceptable administrative waste and increase litigation without promoting preservation goals.

India passed the Biological Diversity Act in 2002 for the conservation of biological diversity in India. The Act also provides mechanisms for equitable sharing of benefits from the use of traditional biological resources and knowledge.

Analytical Limits

Taxonomic and Size Relationships

Less than 1% of all species that have been described have been studied beyond simply noting their existence. The vast majority of Earth's species are microbial. Contemporary biodiversity physics is "firmly fixated on the visible [macroscopic] world". For example, microbial life is metabolically and environmentally more diverse than multicellular life. "On the tree of life, based on analyses of small-subunit ribo-somal RNA, visible life consists of barely noticeable twigs. The inverse relationship of size and population recurs higher on the evolutionary ladder—"to a first approxima-tion, all multicellular species on Earth are insects". Insect extinction rates are highsupporting the Holocene extinction hypothesis.

Plant Breeders' Rights

Plant breeders' rights (PBR), also known as plant variety rights (PVR), are rights granted to the breeder of a new variety of plant that give the breeder exclusive control over the propagating material (including seed, cuttings, divisions, tissue culture) and harvested material (cut flowers, fruit, foliage) of a new variety for a number of years.

With these rights, the breeder can choose to become the exclusive marketer of the variety, or to license the variety to others. In order to qualify for these exclusive rights, a variety must be new, distinct, uniform and stable. A variety is:

- new if it has not been commercialized for more than one year in the country of protection;
- distinct if it differs from all other known varieties by one or more important botanical characteristics, such as height, maturity, color, etc.;
- uniform if the plant characteristics are consistent from plant to plant within the variety;
- stable if the plant characteristics are genetically fixed and therefore remain the

same from generation to generation, or after a cycle of reproduction in the case of hybrid varieties.

The breeder must also give the variety an acceptable "denomination", which becomes its generic name and must be used by anyone who markets the variety.

Typically, plant variety rights are granted by national offices, after examination. Seed is submitted to the plant variety office, who grow it for one or more seasons, to check that it is distinct, stable, and uniform. If these tests are passed, exclusive rights are granted for a specified period (typically 20/25 years (or 25/30 years, for trees and vines). Annual renewal fees are required to maintain the rights.

Breeders can bring suit to enforce their rights and can recover damages for infringement. Plant breeders' rights contain exemptions from infringement that are not recognized under patent law. Commonly, there is an exemption for farm-saved seed. Farmers may store the production in their own bins for their own use as seed, but this does not necessarily extend to brown-bag sales of seed. Further sales for propagation purposes are not allowed without the written approval of the breeder. There is also a breeders' exemption (research exemption in the 1991 Act) that allows breeders to use protected varieties as sources of initial variation to create new varieties of plants (1978 Act), or for other experimental purposes (1991 Act). There is also a provision for compulsory licensing to assure public access to protected varieties if the national interest requires it and the breeder is unable to meet the demand.

There is tension over the relationship between patent rights and plant breeder's rights. There has been litigation in Australia, the United States, and Canada over the overlap between such rights. Each of these cases was decided on the principle that patents and plant breeders' rights were overlapping and not mutually exclusive. Thus, the exemptions from infringement of plant breeders' rights, such as the saved seed exemption, do not create corresponding exemptions from infringement of the patents covering the same plants. Likewise, acts that infringe the plant breeders' rights, such as exportation of the variety, would not necessarily infringe a patent on the variety, which only allows the patent owner to prohibit making, using or selling the patented invention.

International Rights

In 1957, in France negotiations took place concerned with the protection of new varieties. This led to the creation of the Union Internationale pour la Protection des Obtentions Végétales (UPOV) and adoption of the first text of the International Convention for the Protection of New Varieties of Plants (UPOV Convention) in 1961. The purpose of the Convention was to ensure that the member states party to the Convention acknowledge the achievements of breeders of new plant varieties by making available to them an exclusive property right, on the basis of a set of uniform and clearly defined principles.

The Convention was revised in Geneva in 1972, 1978 and 1991. Both the 1978 and the 1991 Acts set out a minimum scope of protection and offer member States the possibility of taking national circumstances into account in their legislation. Under the 1978 Act, the minimum scope of the plant breeder's right requires that the holder's prior authorisation is necessary for the production for purposes of commercial marketing, the offering for sale and the marketing of propagating material of the protected variety. The 1991 Act contains more detailed provisions defining the acts concerning propagating material in relation to which the holder's authorisation is required. The breeder's authorisation is also required in relation to any of the specified acts done with harvested material of the variety, unless the breeder has had reasonable opportunity to exercise their right in relation to the propagating material, or if not doing so could constitute an "Omega Threat" situation. Under that provision, for example, a flower breeder who protects their variety in the Netherlands could block importation of cut flowers of that variety into the Netherlands from Egypt, which does not grant plant breeders' rights, because the breeder had no opportunity to exercise any rights in Egypt. Member countries also have the option to require the breeder's authorization with respect to the specified acts as applied to products directly obtained from the harvested material (such as flour or oil from grain, or juice from fruit), unless the breeder has had reasonable opportunity to exercise their right in relation to the harvested material.

The UPOV Convention also establishes a multilateral system of *national treatment*, under which citizens of any member state are treated as citizens of all member states for the purpose of obtaining plant breeders rights. It also sets up a multilateral priority filing system, under which an application for protection filed in one member state establishes a filing date for applications filed in all other member states within one year of that original filing date. This allows a breeder to file in any one member country within the one-year period required to preserve the novelty of their variety, and the novelty of the variety will still be recognized when the filing is done in other member countries within one year of the original filing date. However, if the applicant does not wish to make use of priority filing, he or she has four years in which to apply in all other member states, excepting the USA, for all species except tree and vine species in which case he or she has six years to make application. More information can be obtained in Article 10 (1) (b) of Council Regulation (EC) No. 2100/94 of 27 July 2004. The trigger to start the four- or six-year period is not actually the date on which the first filing is made but the date on which the variety was first commercialised.

The UPOV Convention is not self-executing. Each member state must adopt legislation consistent with the requirements of the convention and submit that legislation to the UPOV Secretariat for review and approval by the UPOV Council, which consists of all the UPOV member states acting in committee. In compliance with these treaty obligations, the United Kingdom enacted the Plant Variety and Seeds Act 1964. Similar legislation was passed in the Netherlands, Denmark, Germany, and New Zealand. In 1970 the United States followed the lead of seventeen Western European nations and passed the Plant Variety Protection Act 1970 (US). This legislation provided protection to developers of novel, sexually reproduced plants. However, the United States originally acceded to the UPOV Convention on the basis of the Plant Patent Act and did not bring the PVP Act into compliance with UPOV requirements until 1984 when the Commissioner of Plant Variety Protection promulgated rules to do so. Since the 1980s, the US Patent Office has granted patents on plants, including plant varieties this provides a second way of protecting plant varieties in the USA. Australia passed the Plant Variety Protection Act 1987 (Cth) and the Plant Breeders Rights Act 1994 (Cth). Australian patent law also permits the patenting of plant varieties. In total, 65 countries have signed the UPOV Convention and adopted plant breeders' rights legislation consistent with the requirements of the convention.

The WTO's Agreement on Trade-Related Aspects of Intellectual Property Rights (TRIPs) requires member states to provide protection for plant varieties either by patents or by an effective *sui generis* (stand alone) system, or a combination of the two. Most countries meet this requirement through UPOV Convention-compliant legislation. India has adopted a plant breeders' rights law that has been rejected by the UPOV Council as not meeting the requirements of the treaty.

References

- Whitney, Ellie; Rolfes, Sharon Rady (2013). Understanding Nutrition (13 ed.). Wadsworth, Cengage Learning. pp. 667, 670. ISBN 978-1133587521.
- Payne-Palacio, June R.; Canter, Deborah D. (2014). The Profession of Dietetics. Jones & Bartlett Learning. pp. 3, 4. ISBN 978-1-284-02608-5.
- Silberberg, Martin S. (2009). Chemistry: The Molecular Nature of Matter and Change (5 ed.). McGraw-Hill. p. 44. ISBN 978-0-07-304859-8.
- Nelson, D. L.; Cox, M. M. (2000). Lehninger Principles of Biochemistry (3rd ed.). New York: Worth Publishing. ISBN 1-57259-153-6.
- Corbridge, D. E. C. (1995). Phosphorus: An Outline of its Chemistry, Biochemistry, and Technology (5th ed.). Amsterdam: Elsevier. ISBN 0-444-89307-5.
- Lippard, S. J.; Berg, J. M. (1994). Principles of Bioinorganic Chemistry. Mill Valley, CA: University Science Books. ISBN 0935702725.
- Stearns, Beverly Peterson; Stearns, S. C.; Stearns, Stephen C. (2000). Watching, from the Edge of Extinction. Yale University Press. p. 1921. ISBN 978-0-300-08469-6. Retrieved 2014-12-27.
- Morand, Serge; Krasnov, Boris R. (1 September 2010). The Biogeography of Host-Parasite Interactions. Oxford University Press. pp. 93–94. ISBN 978-0-19-956135-3. Retrieved 28 June 2011.
- Vandermeer, John H. (2011). The Ecology of Agroecosystems. Jones & Bartlett Learning. ISBN 978-0-7637-7153-9.
- Chivian, Eric, ed. (15 May 2008). Sustaining Life: How Human Health Depends on Biodiversity. OUP USA. ISBN 978-0-19-517509-7.

5

Essential Fields of Plant Breeding

The associated fields of plant breeding dealt within this chapter are cultivated plant taxonomy and organic farming. The study of the theory and practice of the science that identifies, describes and classifies those plants whose origin is primarily due to intentional human activity is known as cultivated plant taxonomy. The topics discussed in the chapter are of great importance to broaden the existing knowledge on plant breeding.

Cultivated Plant Taxonomy

Cultivated plant taxonomy is the study of the theory and practice of the science that identifies, describes, classifies, and names cultigens—those plants whose origin or selection is primarily due to intentional human activity. Cultivated plant taxonomists do, however, work with all kinds of plants in cultivation.

Some of the traditional tools of cultivated plant taxonomy including: microscope, camera, flowers and book to assist identification.

Cultivated plant taxonomy is one part of the study of horticultural botany which is mostly carried out in botanical gardens, large nurseries, universities, or government departments. Areas of special interest for the cultivated plant taxonomist include: searching for and recording new plants suitable for cultivation (plant hunting); communicating with and advising the general public on matters concerning the classification and nomenclature of cultivated plants and carrying out original research on these topics; describing the cultivated plants of particular regions (horticultural floras); maintaining databases, herbaria and other information about cultivated plants.

Much of the work of the cultivated plant taxonomist is concerned with the naming of plants as prescribed by two plant nomenclatural Codes. The provisions of the *International Code of Nomenclature for algae, fungi, and plants* (*Botanical Code*) serve primarily scientific ends and the objectives of the scientific community, while those of the *International Code of Nomenclature for Cultivated Plants* (*Cultivated Plant Code*) are designed to serve both scientific and utilitarian ends by making provision for the names of plants used in commerce — the cultigens that have arisen in agriculture, forestry and horticulture. These names, sometimes called variety names, are not in Latin but are added onto the scientific Latin names, and they assist communication among the community of foresters, farmers and horticulturists.

The history of cultivated plant taxonomy can be traced from the first plant selections that occurred during the agrarian Neolithic Revolution to the first recorded naming of human plant selections by the Romans. The naming and classification of cultigens followed a similar path to that of all plants until the establishment of the first *Cultivated Plant Code* in 1953 which formally established the cultigen classification category of cultivar. Since that time the classification and naming of cultigens has followed its own path.

Distinctive Characteristics

Cultivated plant taxonomy has been distinguished from the taxonomy of other plants in at least five ways. Firstly, there is a distinction made according to *where the plants are growing* — that is, whether they are wild or cultivated. This is alluded to by the *Cultivated Plant Code* which specifies in its title that it is dealing with *cultivated plants*. Secondly, a distinction is made according to *how the plants originated*. This is indicated in Principle 2 of the *Cultivated Plant Code* which defines the scope of the Code as "*... plants whose origin or selection is primarily due to the intentional actions of mankind*" — plants that have evolved under natural selection with human assistance.Thirdly, cultivated plant taxonomy is concerned with plant variation that requires the use of special classification categories that do not conform with the hierarchy of ranks implicit in the *Botanical Code*, these categories being the cultivar, Group and grex (which are only loosely equivalent to ranks in the *Botanical Code*). This feature is also referred to in the Preamble to the *Cultivated Plant Code* which states that "*The purpose of giving a name to a taxon is not to indicate its characters or history, but to supply a means of referring to it and to indicate to which category it is assigned.*" Fourthly, cultivated plant taxonomy serves a particular community of people: the *Botanical Code* focuses on the needs of plant taxonomists as they attempt to maintain order and stability for the scientific names of all plants, while the *Cultivated Plant Code* caters for the needs of people requiring names for plants used in the commercial world of agriculture, forestry and horticulture. Finally, the difference between cultivated plant taxonomy and the taxonomy of other plants has been attributed to the purpose for which the taxonomy has been devised, it being plant-centred in the *Botanical Code* and human-centred in the *Cultivated Plant Code*.

Scientific and Anthropocentric Classification

The key activities of cultivated plant taxonomy relate to classification (taxonomy) and naming (nomenclature). The rules associated with naming plants are separate from the methods, principles or purposes of classification, except that the units of classification, the taxa, are placed in a nested hierarchy of ranks – like species within genera, and genera within families. There are three classification categories used in the *Cultivated Plant Code*, the cultivar and the Group and the grex, but they are only loosely equivalent to ranks in the *Botanical Code*.

From the time of the ancient world, at least, plants have been classified in two ways. On the one hand there is the detached academic, philosophical or scientific interest in plants themselves: this groups plants by their relationship to one another according to their similarities and differences in structure and function. Then there is the practical, utilitarian or anthropocentric interest which groups plants according to their human use. Cultivated plant taxonomy is concerned with the special classification categories needed for the plants of agriculture, horticulture and forestry as regulated by the *Cultivated Plant Code*. This Code serves not only the scientific interests of formal nomenclature, it also caters for the special utilitarian needs of people dealing with the plants of commerce. Those cultigens given names governed by the *Cultivated Plant Code* fulfill three criteria: they have special features considered of sufficient importance to warrant a name; the special features are the result of deliberate human breeding or selection and are not found in wild populations (except in rare cases where the special features represent desirable part of natural variation found in wild populations that is not covered by a scientific name); it is possible to perpetuate the desirable features by propagation in cultivation.

The terms *cultigen* and *cultivar* may be confused with each other. *Cultigen* is a general-purpose term for plants that have been deliberately altered or specially selected by humans, while cultivar is a formal classification category. Cultigens include not only plants with cultivar names but also those with names in the classification categories of grex and Group. The *Cultivated Plant Code* points out that cultigens are: *deliberately selected plants that may have arisen by intentional or accidental hybridization in cultivation, by selection from existing cultivated stocks, or from variants within wild populations that are maintained as recognizable entities solely by continued propagation.* Included within the group of plants known as cultigens are genetically modified plants, plants with binomial Latin names that are the result of ancient human selection, and any plants that have been altered by humans but which have not been given formal names. In practice most cultigens are cultivars.

The following account of the historical development of cultivated plant taxonomy traces the way cultigens have arisen and been incorporated into botanical science; it also demonstrates how two approaches to plant nomenclature and classification have led to the present-day *International Code of Nomenclature for algae, fungi, and plants* and *International Code of Nomenclature for Cultivated Plants.*

Historical Development

The history of cultigen nomenclature has been discussed by William T. Stearn and Brandenberg, Hetterscheid and Berg. It has also been examined from a botanical perspective and from the origin of the *Cultivated Plant Code* in 1953 until 2004.

The early development of cultigen taxonomy follows that of plant taxonomy in general as the early listing and documentation of plants made little distinction between those that were anthropogenic and those that were natural wild kinds. Formal botanical nomenclature and classification evolved from the simple binomial system of folk taxonomy and it was not until the mid-19th century that the nomenclatural path of cultigens began to diverge from mainstream plant taxonomy.

10,000 to 400 BCE – Plant Domestication

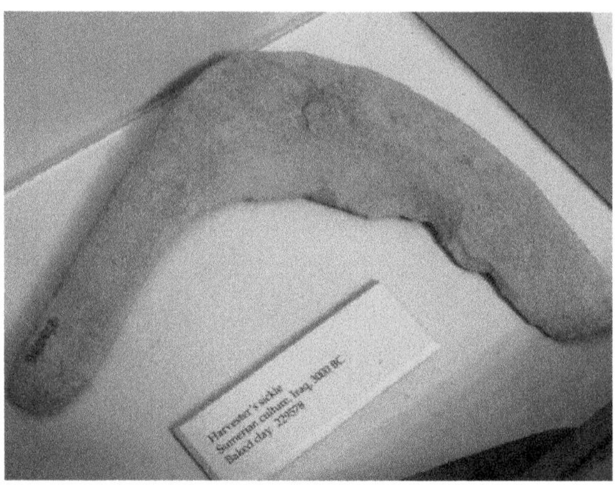

Sumerian harvester's sickle made from baked clay and dated to about 3000 BCE.

William T. Stearn (1911–2001), taxonomic botanist, classical scholar and author of the book *Botanical Latin* has commented that "cultivated plants [cultigens] are mankind's most vital and precious heritage from remote antiquity". Cultigens of our most common economic plants probably date back to the first settled communities of the Neolithic Revolution 10,000 to 12,000 years ago although their exact time and place of true origin will probably remain a mystery. In the Western world among the first cultigens would have been selections of the cereals wheat and barley that arose in the early settlements of the Fertile Crescent (the fertile river valleys of the Nile, Tigris and Euphrates) in the Western Mediterranean. Food plant selections would also have been made in the ten or so other centres of settlement that occurred around the world at this time. Confining crops to local areas gave rise to landraces (selections that are highly adapted to local conditions) although these are now largely replaced by modern cultivars. Cuttings are an extremely effective way of perpetuating desirable characters, especially of woody plants like grapes, figs and olives so it is not surprising that these are among the first known plant selections perpetuated in cultivation in the West. Migrating people would take their plant seeds and cuttings with them; there is evidence of early Fertile Crescent cereal cultigens being transferred from Western Asia to surrounding lands.

400 BCE to 1400 – The Ancient World: Greco-roman Influence to the Middle Ages

Aristotle (384–322 BCE)

As early as the 5th century BCE the Greek philosopher Hippo expressed the opinion that cultigens (as we call them now) were produced from wild plants as the result of the care bestowed on them by man, a revolutionary view at a time when they were regarded as the special creation and gift of the gods. In devising ways of classifying organisms the philosopher Aristotle (384–322 BCE) established the important idea of a *fundamentum divisionis* – the principle that groups can be progressively subdivided. This has been assumed in biological classification ever since and is congruent with the relatively recent idea of evolution as descent with modification. All biological classification follows this principle of groups within groups, known as a nested hierarchy, but this form of classification does not necessarily presuppose evolution.

Theophrastus (371–286 BCE)

The earliest scientific (rather than utilitarian) approach to plants is attributed to Aristotle's student Theophrastus (371–286 BCE), known as the "father of botany". In his *Enquiry into Plants* Theophrastus described 480 kinds of plant, dividing the plant kingdom into trees, shrubs, undershrubs and herbs with further subdivision into wild and cultivated, flowering and non-flowering, deciduous or evergreen.

The utilitarian approach, classifying plants according to their medicinal properties, is exemplified by the work of Roman nobleman, scientist and historian, Pliny the Elder (29–79 CE) author of *Naturalis historiae.* "Cultivars" listed here are named after people, places or special plant characteristics. Most notable is the work of Dioscorides (ca.40–ca.90 CE) a Greek doctor who worked with the Roman army. His five-volume *Materia Medica* was a forerunner of the herbal which led to the modern pharmacopoeia. This work was endlessly plagiarised by later herbals including those printed between about 1470 and 1670 CE: it listed 600 to 1000 different kinds of plants including the cultigens Gallica, Centifolia, the rose of uncertain origin known as Alba and other rose cultivars grown by the Romans.

Pliny the Elder (29–79 CE)

The first record of a named cultigen occurs in *De Agri Cultura*. written about 160 BCE by Roman statesman Cato the Elder (234–149 BCE) in a list that includes 120 kinds (cultivars) of figs, grapes, apples and olives. The names are presented in a way that implies that they would have been familiar to fellow Romans. The "cultivar" names were mostly of one word and denoted the provenance of the cultivar (the geographical origin of the place where the plant selections were made). Writers up to the 15th century added little to this early work. In the Middle Ages the book of hours, early herbals, illuminated manuscripts and economic records indicate that plants grown by the Romans found their way into monastery gardens. For example, in 827 CE the following herbs were mentioned in the poem *Hortulus* by Walafrid Strabo as growing in the monastery garden of St Gallen in Switzerland: sage, rue, southernwood, wormwood, horehound, fennel, German iris, lovage, chervil, Madonna lily, opium poppy, clary, mint, betony, agrimony, catmint, radish, gallica rose, bottle gourd and melon. It seems likely that aromatic and culinary herbs were quite widespread and similar lists of plants occur in records of plants grown in Villa gardens at the time of Charlemagne (742–814 CE).

1400 to 1700 – Renaissance, Imperial Expansion, Herbals

Caspar Bauhin (1550–1624)

The revival of learning during the Renaissance reinvigorated the study of plants and their classification. From about 1400 CE European expansion established Latin as the common language of scholars and it was adopted for biological nomenclature. Then, from about 1500 CE, the publication of herbals (books often illustrated with woodcuts describing the appearance, medicinal properties, and other characteristics of plants used in herbal medicine) extended the formal documentation of plants and by the late 16th century the number of different plant kinds described in Europe had risen to about 4,000. In 1623 Gaspard Bauhin published his *Pinax theatre botanici* an attempt at a comprehensive compilation of all plants known at that time: it included about 6000 kinds. The combined works of a German physician and botanist Valerius Cordus (1515–1544 CE) which were published in 1562 included many named "cultivars" including 30 apples and 49 pears, presumably local German selections. English herbalist John Parkinson's *Paradisi in Sole* ... (1629) lists 57 apple "cultivars", 62 pears, 61 plums, 35 cherries and 22 peaches.

With increasing trade in economic and medicinal plants the need for a more comprehensive classification system increased. Up to about 1650 CE plants had been grouped either alphabetically or according to utilitarian folk taxonomy – by their medicinal uses or whether they were trees, shrubs or herbs. Between 1650 and 1700 CE there was a move from the utilitarian back to a scientific natural classification based on the characters of the plants themselves.

1700 to 1750 – Dawn of Scientific Classification

In 1700 French botanist J.P. de Tournefort although still using the broad groupings of "trees" and "herbs" for flowering plants, began to use flower characteristics as distinguishing features and, most importantly, provided a clear definition of the genus as a basic unit of classification. In *Institutiones Rei Herbariae* he listed about 10,000 different plants, which he called species, organised into 698 genera with illustrations. The establishment of this precursor of scientific classification vastly improved the organisation of plant variation into approximately equivalent groups or ranks and many of his genera were later taken up by Carl Linnaeus. There was still at this time no common agreement on the way to present plant names so they ranged in length from one word to lengthy descriptive sentences. As the number of recorded plants increased this naming system became more unwieldy.

In England the tradition of documenting garden plants was established long before Linnaeus' *Species Plantarum* starting with the herbals, but the most prominent early chronicler was Philip Miller (1691–1771) who was a master gardener in charge of the Chelsea Physic Garden in London from 1722 to 1770. New plants were coming into Western Europe from southern Europe and the overseas colonies of the Dutch, British and French. These new plants came largely to the botanic gardens of Amsterdam, Leiden, Chelsea and Paris and they needed recording. In 1724 Miller produced a two-volume compendium of garden plants called *The Gardeners and Florists Dictionary or a complete System of Horticulture*. The first edition was in 1724, subsequently revised and enlarged until the last and 8th edition in 1768 by which time he had adopted Linnaean binomials. For a while this publication was taken as the starting point for "horticultural" nomenclature equivalent to Linnaeus' *Species Plantarum* which is now taken as the starting point for botanical nomenclature in general. Miller's Dictionary was the first of many English horticultural compendia whose history has been traced by William Stearn.

1750 to 1800 – Linnaeus and Binomial Nomenclature

In the early 18th century colonial expansion and exploration created a demand for the description of thousands of new organisms. This highlighted difficulties in communication about plants, the replication of their descriptions, and the importance of an agreed way of presenting, publishing and applying their names.

Carl Linnaeus (1707–1778) who established the binomial system of plant nomenclature.

It was the Swedish botanist Carl Linnaeus who finally put order into this situation as he attempted to name all the known organisms of his day. In 1735 his *Systema Naturae*, which included animals (the tenth edition became the starting point for zoological nomenclature) was followed by *Critica Botanica* in 1737, and *Philosophia Botanica* in 1751. But it was his most comprehensive work on plants, the 1753 publication *Species Plantarum* that formalised the name of a genus with a single epithet to form the name of a species as two words, the binomial thus making secure the biological system of binomial nomenclature. In these works Linnaeus used a third name as a variety within a species. These varieties included both wild and horticultural variants. The horticultural varieties were still written in Latin and some have persisted to this day.

Linnaeus had very definite and uncomplimentary views about cultigens, regarding them as inferior plants for the amusement of those people he disparagingly called anthophiles (flower-lovers); these were plants not deserving the attention of serious botanists. His views revealed both his prejudice, his stance on special creation, and his recognition of the difficulties entailed in cultivated plant taxonomy:

> *"anthophiles ... practice a floral science all their own, grasped only by their devotees; no botanist in his senses will enlist in their camp."*

> *"All the species recognized by Botanists came forth from the Almighty Creator's hand, and the number now and always will be exactly the same, while every day new and different florist's species arise from the true species recognized by botanists, and when they have arisen they eventually revert to their original forms. Accordingly to the former have been assigned by Nature fixed limits, beyond which they cannot go: while the latter display without end the infinite sport of Nature."*

> *"... botany has been overborne by the system of varieties for long enough ... few, if any, agree as to what constitutes a species, or what a variety; ... I wish the system of varieties were entirely excluded from Botany and turned over entirely to the Anthophiles, since it causes nothing but ambiguities, errors, dead weight and vanity ..."*

1800 to 1900 – Global Plant Trade

CAROLI LINNÆI
S:Æ R:GIÆ M:TIS SVECIÆ ARCHIATRI; MEDIC. & BOTAN. PROFESS. UPSAL; EQUITIS AUR. DE STELLA POLARI; nec non ACAD. IMPER. MONSPEL. BEROL. TOLOS. UPSAL. STOCKH. SOC. & PARIS. CORESP.

SPECIES PLANTARUM,

EXHIBENTES PLANTAS RITE COGNITAS, AD GENERA RELATAS, CUM DIFFERENTIIS SPECIFICIS, NOMINIBUS TRIVIALIBUS, SYNONYMIS SELECTIS, LOCIS NATALIBUS, SECUNDUM SYSTEMA SEXUALE DIGESTAS.

TOMUS I.

Cum Privilegio S. R. Mtis Sueciæ & S. R. Mtis Poloniæ ac Electoris Saxon.

HOLMIÆ, IMPENSIS LAURENTII SALVII. 1753.

C. Appelgren

Linnaeus's *Species Plantarum* of 1753, his catalogue of all the world's plants known to European science.

The natural distribution of plants across the world has determined when and where cultigens have been produced. The botanical and horticultural collection of economically important plants, including ornamentals, was based in Europe. Although economic herbs and spices had a long history in trade, and there are good records of cultivar distribution by the Romans, European botanical and horticultural exploration rapidly increased in the 19th century with the colonial expansion taking place at the time. New plants were brought back to Europe while, at the same time, valuable economic plants, including those from the tropics, were distributed among the colonies. This plant trade has provided the common global heritage of economic and ornamental cultigens that we use today and which formed the stock for modern plant selection, breeding, and genetic engineering. The plant exchange that occurred as a result of European trade can be divided into several phases:

- to 1560 mostly within Europe
- 1560–1620 Near East (esp. bulbous plants from Turkey – "tulipomania")
- 1620–1686 Canada and Virginia herbaceous plants
- 1687–1772 Cape of South Africa
- 1687–1772 North American trees and shrubs
- 1772–1820 Australia, Tasmania, New Zealand

- 1820–1900 Tropical glasshouse plants; hardy Japanese plants
- 1900–1930 West China
- 1930 Intensive breeding and selection programs

1900 to 1950 – The Botanical Code and Cultigen Nomenclature

As the community of people dealing with the cultigens of commerce grew so, once again, the divergence between taxonomy serving scientific purposes and utilitarian taxonomy meeting human needs re-emerged. In 1865 German botanist Karl Koch, who became General Secretary of the Berlin Horticultural Society, expressed resentment at the continued use of Latin for cultigen names. Many proposals to deal with this were made, perhaps the most prominent being the *Lois de la nomenclature botanique* submitted in 1867 to the fourth Horticultural and Botanical Congress by Swiss botanist Alphonse de Candolle who, in Article 40 stated:

"Seedlings, half-breeds (métis) of unknown origin or sports should receive from horticulturists fancy names (noms de fantaisie) in common language, as distinct as possible from the Latin names of species or varieties."

Outstanding American cultivated plant taxonomist Liberty Hyde Bailey (1858–1954).

This Article, making provision for the cultigens of horticultural nomenclature was to remain in the *Botanical Code* (with a minor amendment in 1935 suggesting the use of the letter 'c' before the horticultural name and antedating formal recognition of the cultivar) through 1906, 1912 and 1935 until the separation, in 1953, of the *Horticultural Code*, precursor to the *International Code of Nomenclature for Cultivated Plants* (*Cultivated Plant Code*). In 1900 there was the first International Botanical Congress and in 1905 at the second Congress in Vienna an agreed set of nomenclatural rules was established, the *Vienna Rules*, which became known from then on as the *International*

Code of Botanical Nomenclature (now the *International Code of Nomenclature for algae, fungi, and plants*). After World War II the responsibility for the *Botanical Code* was taken up by the International Association for Plant Taxonomy and meetings to discuss revisions are held at six-yearly intervals, the latest being in 2005

In horticulture at this time there existed all the problems that had confronted botanists in the 19th century – a plethora of names of various length, written and published in many languages with much duplication. The period between 1867 and 1953 was an uneasy time in which American horticulturists and other groups in Europe, such as the specialist orchid community, made attempts to put order into this chaos within their particular group of interest and devising their own rules for naming the plants of commerce. Friedrich Alefeld (1820–1872), who used Latin variety names, in a monographic study of beans, lentils and other legumes distinguished three infraspecific taxonomic categories: *Unterart* (subspecies), *Varietäten Gruppe* and *Kultur-Varietät*, all with Latin names. In doing this he was probably laying the ground for the later establishment of the cultigen classification categories cultivar and Group. In conjunction with the Brussels International Botanical Congress of 1910 there was an International Horticultural Congress having a horticultural nomenclature component.

As a result of general dissatisfaction and a submission from the Royal Horticultural Society the *Règles de Nomenclature Horticole* was established. The use of simple descriptive Latin names (e.g. *compactus, nanus, prostratus*) for horticultural variants was accepted and so too were names in the local language – which were not to be translated and should preferably consist of one word and a maximum of three. This first *Horticultural Code* consisted of 16 Articles. With the intercession of a World War I it was not until the *9th Horticultural Congress* in London in 1930 that the rules of a Horticulture Nomenclature Committee were agreed and added as an appendix to the 1935 *Botanical Code*. The rules established in 1935 were accepted but needed to be extended to include the cultigens of agriculture and forestry, but it was only a result of discussions at the 1950 International Botanical Congress in Stockholm and the 18th International Horticultural Congress in London in 1952 the first *International Code of Nomenclature for Cultivated Plants* was published in 1953. The American horticultural botanist Liberty Hyde Bailey was responsible for coining the word cultigen in 1918 and cultivar in 1923, the word cultivar only coming into general circulation with the new Code of 1953. The use of these two terms belies the multitude of classification terms and categories that had been suggested as designations for cultigens.

1953 – The International Code of Nomenclature for Cultivated Plants

The first *Cultivated Plant Code* (Wageningen), which was published in 1953, has been followed by eight subsequent editions – in 1958 (Utrecht), 1961 (update of 1958), 1969 (Edinburgh), 1980 (Seattle), 1995 (Edinburgh), 2004 (Toronto) and 2009 (Wageningen).

Following the structure of the *Botanical Code* the *Cultivated Plant Code* is set out in the form of an initial set of Principles followed by Rules and Recommendations that are subdivided into Articles. Amendments to the *Cultivated Plant Code* are prompted by international symposia for cultivated plant taxonomy which allow for rulings made by the *International Commission on the Nomenclature of Cultivated Plants*. Each new *Cultivated Plant Code* includes a summary of the changes made to the previous version and these have also been summarised for the period 1953 to 1995.

International Association for Cultivated Plant Taxonomy

Wisley is one of the Royal Horticultural Society's flagship gardens and a focus for cultivated plant taxonomy.

Recent concerns have focused on international communication on cultivated plant taxonomy, organisation of international symposia, and general communication on topics of interest. In 1988 a *Horticultural Taxonomy Group* (*Hortax*)[nb 1] was formed in the UK and a parallel organisation, the *Nomenclature and Registration Working Group* of the *Vaste Keurings Commissie* in the Netherlands. One development promoting discussion was the newsletter *Hortax News* which was superseded in February 2006 by the first issue of *Hanburyana*, a journal produced by the Royal Horticultural Society in London and dedicated to horticultural taxonomy. This filled a gap left when the American journal *Baileya* ceased publication in the early 1990s. Another development was the launch, in 2007, at the Sixth Symposium on the Taxonomy of Cultivated Plants at Wageningen of the International Association for Cultivated Plant Taxonomy. *Hortax* also publishes *Plant Names: A Guide for Horticulturists, Nurserymen, Gardeners and Students*.

Presenting Cultigen Names

Most cultigens have names consisting of a Latin name that is governed by the *International Code of Nomenclature for algae, fungi, and plants* e.g. *Malus domestica*, to which is added a cultigen epithet, enclosed in single quotes e.g. *Malus domestica* 'Granny Smith'. The formation and use of the three classification categories (ranks) used for cultigens, the cultivar, Group and grex, is regulated by the ICNCP. Examples of acceptable ways to present cultigen names are given below:

Prunus serrata Sato-zakura Group

Prunus serrata (Sato-zakura Group) 'Ojochin'

Prunus 'Ojochin'

Flowering cherry 'Ojochin'

Contemporary Issues

Chelsea Physic Garden, summer 2006

Current challenges for cultivated plant taxonomists include: the use of large plant name databases; ways of dealing with the use of non-scientific names in commerce (known as trade designations), especially for plant labels in nurseries; intellectual property and plants; adapting modern technology, in particular molecular techniques, to the creation and identification of cultivars; maintaining germplasm collections of cultivars, including herbaria; the recording and registration of cultivars.

The ways in which the plant variation resulting from human activity is named and classified remains contentious. The replacement of the expression "cultivated plant" with the word "cultigen" is not universally accepted. The debate continues concerning the notions of ranks and taxa as applied to cultigens. Is it appropriate to call the highly modified transgenic products of human artificial selection "taxa" in the same way we do for the products of natural selection in the wild? To overcome this difficulty the term *culton* (pl. *culta*) has been suggested to replace the word taxon when speaking about cultigens.

Then, most "wild" plants fit neatly into the nested hierarchy of ranks used in Linnaean classification (species into genera, genera into families etc.) which aligns with Darwinian descent with modification. Choosing classification categories for cultigens is not clear-cut. Included among cultigens are: simple selections taken from plants in the wild or in cultivation; artificial hybrids produced both by accident and intention; plants produced by genetic engineering; clonal material reproduced by cuttings, grafting, budding, layering etc.; graft-chimaeras; selections from the wild; ancient selections of crops that date back thousands of years; selections of aberrant growth such as witches brooms; the results of deliberate repeatable single crosses between two pure lines to produce plants of a particular general appearance that is desirable for horticulture, but which are not genetically identical. The question remains as to whether the classification categories of cultivar, Group and grex are the most appropriate and efficient way to deal with this broad range of plant variation.

Organic Farming

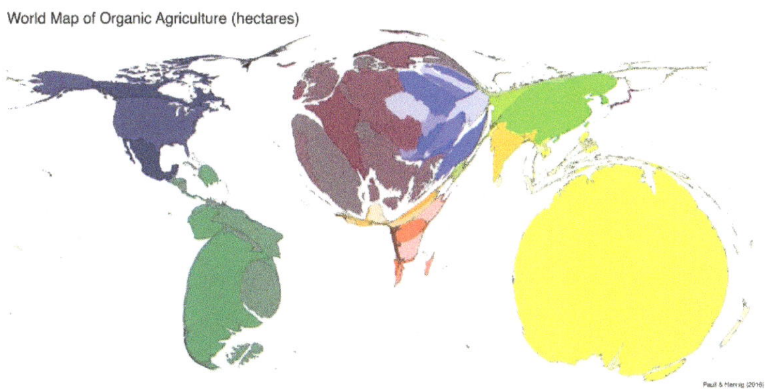

World map of organic agriculture (hectares)

Vegetables from ecological farming.

Organic farming is an alternative agricultural system which originated early in the 20th century in reaction to rapidly changing farming practices. Organic agriculture continues to be developed by various organic agriculture organizations today. It relies on fertilizers of organic origin such as compost, manure, green manure, and bone meal and places emphasis on techniques such as crop rotation and companion planting. Biological pest control, mixed cropping and the fostering of insect predators are encouraged. In general, organic standards are designed to allow the use of naturally occurring substances while prohibiting or strictly limiting synthetic substances. For instance, naturally occurring pesticides such as pyrethrin and rotenone are permitted, while synthetic fertilizers and pesticides are generally prohibited. Synthetic substances that are allowed include, for example, copper sulfate, elemental sulfur and Ivermectin. Genetically modified organisms, nanomaterials, human sewage sludge, plant growth regulators, hormones, and antibiotic use in livestock husbandry are prohibited. Reasons for advocation of organic farming include real or perceived advantages in sustainability, openness, self-sufficiency, autonomy/independence, health, food security, and food safety, although the match between perception and reality is continually challenged.

Organic agricultural methods are internationally regulated and legally enforced by many nations, based in large part on the standards set by the International Federation of Organic Agriculture Movements (IFOAM), an international umbrella organization for organic farming organizations established in 1972. Organic agriculture can be defined as:

an integrated farming system that strives for sustainability, the enhancement of soil fertility and biological diversity whilst, with rare exceptions, prohibiting synthetic pesticides, antibiotics, synthetic fertilizers, genetically modified organisms, and growth hormones.

Since 1990 the market for organic food and other products has grown rapidly, reaching $63 billion worldwide in 2012. This demand has driven a similar increase in organically managed farmland that grew from 2001 to 2011 at a compounding rate of 8.9% per annum. As of 2011, approximately 37,000,000 hectares (91,000,000 acres) worldwide were farmed organically, representing approximately 0.9 percent of total world farmland.

History

Agriculture was practiced for thousands of years without the use of artificial chemicals. Artificial fertilizers were first created during the mid-19th century. These early fertilizers were cheap, powerful, and easy to transport in bulk. Similar advances occurred in chemical pesticides in the 1940s, leading to the decade being referred to as the 'pesticide era'. These new agricultural techniques, while beneficial in the short term, had serious longer term side effects such as soil compaction, erosion, and declines in overall soil fertility, along with health concerns about toxic chemicals entering the food supply. In the late 1800s and early 1900s, soil biology scientists began to seek ways to remedy these side effects while still maintaining higher production.

Biodynamic agriculture was the first modern system of agriculture to focus exclusively on organic methods. Its development began in 1924 with a series of eight lectures on agriculture given by Rudolf Steiner. These lectures, the first known presentation of what later came to be known as organic agriculture, were held in response to a request by farmers who noticed degraded soil conditions and a deterioration in the health and quality of crops and livestock resulting from the use of chemical fertilizers. The one hundred eleven attendees, less than half of whom were farmers, came from six countries, primarily Germany and Poland. The lectures were published in November 1924; the first English translation appeared in 1928 as *The Agriculture Course*.

In 1921, Albert Howard and his wife Gabrielle Howard, accomplished botanists, founded an Institute of Plant Industry to improve traditional farming methods in India. Among other things, they brought improved implements and improved animal husbandry methods from their scientific training; then by incorporating aspects of the local traditional methods, developed protocalls for the rotation of crops, erosion prevention techniques, and the systematic use of composts and manures. Stimulated by these experiences of traditional farming, when Albert Howard returned to Britain in the early 1930s he began to promulgate a system of natural agriculture.

In July 1939, Ehrenfried Pfeiffer, the author of the standard work on biodynamic agriculture (*Bio-Dynamic Farming and Gardening*), came to the UK at the invitation of Walter James, 4th Baron Northbourne as a presenter at the Betteshanger Summer School and Conference on Biodynamic Farming at Northbourne's farm in Kent. One of the chief purposes of the conference was to bring together the proponents of various approaches to organic agriculture in order that they might cooperate within a larger movement. Howard attended the conference, where he met Pfeiffer. In the following year, Northbourne published his manifesto of organic farming, *Look to the Land*, in which he coined the term "organic farming." The Betteshanger conference has been described as the 'missing link' between biodynamic agriculture and other forms of organic farming.

In 1940 Howard published his *An Agricultural Testament*. In this book he adopted Northbourne's terminology of "organic farming." Howard's work spread widely, and he became known as the "father of organic farming" for his work in applying scientific knowledge and principles to various traditional and natural methods. In the United States J.I. Rodale, who was keenly interested both in Howard's ideas and in biodynamics, founded in the 1940s both a working organic farm for trials and experimentation, The Rodale Institute, and the Rodale Press to teach and advocate organic methods to the wider public. These became important influences on the spread of organic agriculture. Further work was done by Lady Eve Balfour in the United Kingdom, and many others across the world.

Increasing environmental awareness in the general population in modern times has transformed the originally supply-driven organic movement to a demand-driven one. Premium prices and some government subsidies attracted farmers. In the developing world, many producers farm according to traditional methods that are comparable to organic farming, but not certified, and that may not include the latest scientific advancements in organic agriculture. In other cases, farmers in the developing world have converted to modern organic methods for economic reasons.

Terminology

Biodynamic agriculturists, who based their work on Steiner's spiritually-oriented anthroposophy, used the term "organic" to indicate that a farm should be viewed as a living organism, in the sense of the following quotation:

"An organic farm, properly speaking, is not one that uses certain methods and substances and avoids others; it is a farm whose structure is formed in imitation of the structure of a natural system that has the integrity, the independence and the benign dependence of an organism"

—Wendell Berry, "The Gift of Good Land"

The use of "organic" popularized by Howard and Rodale, on the other hand, refers more narrowly to the use of organic matter derived from plant compost and animal manures to improve the humus content of soils, grounded in the work of early soil scientists who developed what was then called "humus farming." Since the early 1940s the two camps have tended to merge.

Methods

Organic cultivation of mixed vegetables in Capay, California. Note the hedgerow in the background.

"Organic agriculture is a production system that sustains the health of soils, ecosystems and people. It relies on ecological processes, biodiversity and cycles adapted to local conditions, rather than the use of inputs with adverse effects. Organic agriculture combines tradition, innovation and science to benefit the shared environment and promote fair relationships and a good quality of life for all involved..."

—International Federation of Organic Agriculture Movements

Organic farming methods combine scientific knowledge of ecology and modern technology with traditional farming practices based on naturally occurring biological processes. Organic farming methods are studied in the field of agroecology. While conventional agriculture uses synthetic pesticides and water-soluble synthetically purified fertilizers, organic farmers are restricted by regulations to using natural pesticides and fertilizers. An example of a natural pesticide is pyrethrin, which is found naturally in the Chrysanthemum flower. The principal methods of organic farming include crop rotation, green manures and compost, biological pest control, and mechanical cultivation. These measures use the natural environment to enhance agricultural productivity: legumes are planted to fix nitrogen into the soil, natural insect predators are encouraged, crops are rotated to confuse pests and renew soil, and natural materials such as potassium bicarbonate and mulches are used to control disease and weeds. Genetically modified seeds and animals are excluded.

While organic is fundamentally different from conventional because of the use of carbon based fertilizers compared with highly soluble synthetic based fertilizers and biological pest control instead of synthetic pesticides, organic farming and large-scale conventional farming are not entirely mutually exclusive. Many of the methods developed for organic agriculture have been borrowed by more conventional agriculture. For example, Integrated Pest Management is a multifaceted strategy that uses various organic methods of pest control whenever possible, but in conventional farming could include synthetic pesticides only as a last resort.

Crop Diversity

Organic farming encourages Crop diversity. The science of agroecology has revealed the benefits of polyculture (multiple crops in the same space), which is often employed in organic farming. Planting a variety of vegetable crops supports a wider range of beneficial insects, soil microorganisms, and other factors that add up to overall farm health. Crop diversity helps environments thrive and protects species from going extinct.

Soil Management

Organic farming relies heavily on the natural breakdown of organic matter, using techniques like green manure and composting, to replace nutrients taken from the soil by previous crops. This biological process, driven by microorganisms such as mycorrhiza, allows the natural production of nutrients in the soil throughout the growing season, and has been referred to as *feeding the soil to feed the plant*. Organic farming uses a variety of methods to improve soil fertility, including crop rotation, cover cropping, reduced tillage, and application of compost. By reducing tillage, soil is not inverted and exposed to air; less carbon is lost to the atmosphere resulting in more soil organic carbon. This has an added benefit of carbon sequestration, which can reduce green house gases and help reverse climate change.

Plants need nitrogen, phosphorus, and potassium, as well as micronutrients and symbiotic relationships with fungi and other organisms to flourish, but getting enough nitrogen, and particularly synchronization so that plants get enough nitrogen at the right time (when plants need it most), is a challenge for organic farmers. Crop rotation and green manure ("cover crops") help to provide nitrogen through legumes (more precisely, the *Fabaceae* family), which fix nitrogen from the atmosphere through symbiosis with rhizobial bacteria. Intercropping, which is sometimes used for insect and disease control, can also increase soil nutrients, but the competition between the legume and the crop can be problematic and wider spacing between crop rows is required. Crop residues can be ploughed back into the soil, and different plants leave different amounts of nitrogen, potentially aiding synchronization. Organic farmers also use animal manure, certain processed fertilizers such as seed meal and various mineral powders such as rock phosphate and green sand, a naturally occurring form of potash that provides potassium. Together these methods help to control erosion. In some cases pH may need to be amended. Natural pH amendments include lime and sulfur, but in the U.S. some compounds such as iron sulfate, aluminum sulfate, magnesium sulfate, and soluble boron products are allowed in organic farming.

Mixed farms with both livestock and crops can operate as ley farms, whereby the land gathers fertility through growing nitrogen-fixing forage grasses such as white clover or alfalfa and grows cash crops or cereals when fertility is established. Farms without livestock ("stockless") may find it more difficult to maintain soil fertility, and may rely more on external inputs such as imported manure as well as grain legumes and green manures, although grain legumes may fix limited nitrogen because they are harvested. Horticultural farms that grow fruits and vegetables in protected conditions often relay even more on external inputs.

Biological research into soil and soil organisms has proven beneficial to organic farming. Varieties of bacteria and fungi break down chemicals, plant matter and animal waste into productive soil nutrients. In turn, they produce benefits of healthier yields and more productive soil for future crops. Fields with less or no manure display significantly lower yields, due to decreased soil microbe community. Increased manure improves biological activity, providing a healthier, more arable soil system and higher yields.

Weed Management

Organic weed management promotes weed suppression, rather than weed elimination, by enhancing crop competition and phytotoxic effects on weeds. Organic farmers integrate cultural, biological, mechanical, physical and chemical tactics to manage weeds without synthetic herbicides.

Organic standards require rotation of annual crops, meaning that a single crop cannot be grown in the same location without a different, intervening crop. Organic crop rotations frequently include weed-suppressive cover crops and crops with dissimilar life cycles to discourage weeds associated with a particular crop. Research is ongoing to develop organic methods to promote the growth of natural microorganisms that suppress the growth or germination of common weeds.

Other cultural practices used to enhance crop competitiveness and reduce weed pressure include selection of competitive crop varieties, high-density planting, tight row spacing, and late planting into warm soil to encourage rapid crop germination.

Mechanical and physical weed control practices used on organic farms can be broadly grouped as:

- Tillage - Turning the soil between crops to incorporate crop residues and soil amendments; remove existing weed growth and prepare a seedbed for planting; turning soil after seeding to kill weeds, including cultivation of row crops;
- Mowing and cutting - Removing top growth of weeds;
- Flame weeding and thermal weeding - Using heat to kill weeds; and
- Mulching - Blocking weed emergence with organic materials, plastic films, or landscape fabric.

Some critics, citing work published in 1997 by David Pimentel of Cornell University, which described an epidemic of soil erosion worldwide, have raised concerned that tillage contribute to the erosion epidemic. The FAO and other organizations have advocated a 'no-till' approach to both conventional and organic farming, and point out in particular that crop rotation techniques used in organic farming are excellent no-till approaches. A study published in 2005 by Pimentel and colleagues confirmed that 'Crop rotations and cover cropping (green manure) typical of organic agriculture reduce soil erosion, pest problems, and pesticide use.' Some naturally sourced chemicals are allowed for herbicidal use. These include certain formulations of acetic acid (concentrated vinegar), corn gluten meal, and essential oils. A few selective bioherbicides based on fungal pathogens have also been developed. At this time, however, organic herbicides and bioherbicides play a minor role in the organic weed control toolbox.

Weeds can be controlled by grazing. For example, geese have been used successfully to weed a range of organic crops including cotton, strawberries, tobacco, and corn, reviving the practice of keeping cotton patch geese, common in the southern U.S. before the 1950s. Similarly, some rice farmers introduce ducks and fish to wet paddy fields to eat both weeds and insects.

Controlling Other Organisms

Organisms aside from weeds that cause problems on organic farms include arthropods (e.g., insects, mites), nematodes, fungi and bacteria. Organic practices include, but are not limited to:

- encouraging predatory beneficial insects to control pests by serving them nursery plants and/or an alternative habitat, usually in a form of a shelterbelt, hedgerow, or beetle bank;
- encouraging beneficial microorganisms;
- rotating crops to different locations from year to year to interrupt pest reproduction cycles;
- planting companion crops and pest-repelling plants that discourage or divert pests;
- using row covers to protect crops during pest migration periods;
- using biologic pesticides and herbicides
- using stale seed beds to germinate and destroy weeds before planting
- using sanitation to remove pest habitat;
- Using insect traps to monitor and control insect populations.
- Using physical barriers, such as row covers

Chloroxylon is used for Pest Management in Organic Rice Cultivation in Chhattisgarh, India

Examples of predatory beneficial insects include minute pirate bugs, big-eyed bugs, and to a lesser extent ladybugs (which tend to fly away), all of which eat a wide range of pests. Lacewings are also effective, but tend to fly away. Praying mantis tend to move more slowly and eat less heavily. Parasitoid wasps tend to be effective for their selected prey, but like all small insects can be less effective outdoors because the wind controls their movement. Predatory mites are effective for controlling other mites.

Naturally derived insecticides allowed for use on organic farms use include *Bacillus thuringiensis* (a bacterial toxin), pyrethrum (a chrysanthemum extract), spinosad (a bacterial metabolite), neem (a tree extract) and rotenone (a legume root extract). Fewer than 10% of organic farmers use these pesticides regularly; one survey found that only 5.3% of vegetable growers in California use rotenone while 1.7% use pyrethrum. These pesticides are not always more safe or environmentally friendly than synthetic pesticides and can cause harm. The main criterion for organic pesticides is that they are naturally derived, and some naturally derived substances have been controversial. Controversial natural pesticides include rotenone, copper, nicotine sulfate, and pyrethrums Rotenone and pyrethrum are particularly controversial because they work by attacking the nervous system, like most conventional insecticides. Rotenone is extremely toxic to fish and can induce symptoms resembling Parkinson's disease in mammals. Although pyrethrum (natural pyrethrins) is more effective against insects when used with piperonyl butoxide (which retards degradation of the pyrethrins), organic standards generally do not permit use of the latter substance.

Naturally derived fungicides allowed for use on organic farms include the bacteria *Bacillus subtilis* and *Bacillus pumilus*; and the fungus *Trichoderma harzianum*. These are mainly effective for diseases affecting roots. Compost tea contains a mix of beneficial microbes, which may attack or out-compete certain plant pathogens, but variability among formulations and preparation methods may contribute to inconsistent results or even dangerous growth of toxic microbes in compost teas.

Some naturally derived pesticides are not allowed for use on organic farms. These include nicotine sulfate, arsenic, and strychnine.

Synthetic pesticides allowed for use on organic farms include insecticidal soaps and horticultural oils for insect management; and Bordeaux mixture, copper hydroxide and sodium bicarbonate for managing fungi. Copper sulfate and Bordeaux mixture (copper sulfate plus lime), approved for organic use in various jurisdictions, can be more environmentally problematic than some synthetic fungicides dissallowed in organic farming Similar concerns apply to copper hydroxide. Repeated application of copper sulfate or copper hydroxide as a fungicide may eventually result in copper accumulation to toxic levels in soil, and admonitions to avoid excessive accumulations of copper in soil appear in various organic standards and elsewhere. Environmental concerns for several kinds of biota arise at average rates of use of such substances for some crops. In the European Union, where replacement of copper-based fungicides in organic agriculture is a policy priority, research is seeking alternatives for organic production.

Livestock

Raising livestock and poultry, for meat, dairy and eggs, is another traditional farming activity that complements growing. Organic farms attempt to provide animals with natural living conditions and feed. Organic certification verifies that livestock are raised according to the USDA organic regulations throughout their lives. These regulations include the requirement that all animal feed must be certified organic.

For livestock like these healthy cows vaccines play an important part in animal health since antibiotic therapy is prohibited in organic farming

Organic livestock may be, and must be, treated with medicine when they are sick, but drugs cannot be used to promote growth, their feed must be organic, and they must be pastured.

Also, horses and cattle were once a basic farm feature that provided labor, for hauling and plowing, fertility, through recycling of manure, and fuel, in the form of food for farmers and other animals. While today, small growing operations often do not include livestock, domesticated animals are a desirable part of the organic farming equation, especially for true sustainability, the ability of a farm to function as a self-renewing unit.

Genetic Modification

A key characteristic of organic farming is the rejection of genetically engineered plants and animals. On 19 October 1998, participants at IFOAM's 12th Scientific Conference issued the Mar del Plata Declaration, where more than 600 delegates from over 60 countries voted unanimously to exclude the use of genetically modified organisms in food production and agriculture.

Although opposition to the use of any transgenic technologies in organic farming is strong, agricultural researchers Luis Herrera-Estrella and Ariel Alvarez-Morales continue to advocate integration of transgenic technologies into organic farming as the optimal means to sustainable agriculture, particularly in the developing world, as does author and scientist Pamela Ronald, who views this kind of biotechnology as being consistent with organic principles.

Although GMOs are excluded from organic farming, there is concern that the pollen from genetically modified crops is increasingly penetrating organic and heirloom seed stocks, making it difficult, if not impossible, to keep these genomes from entering the organic food supply. Differing regulations among countries limits the availability of

GMOs to certain countries, as described in the article on regulation of the release of genetic modified organisms.

Tools

Organic farmers use a number of traditional farm tools to do farming. Due to the goals of sustainability in organic farming, organic farmers try to minimize their reliance on fossil fuels. In the developing world on small organic farms tools are normally constrained to hand tools and diesel powered water pumps. A recent study evaluated the use of open-source 3-D printers (called RepRaps using a bioplastic polylactic acid (PLA) on organic farms. PLA is a strong biodegradable and recyclable thermoplastic appropriate for a range of representative products in five categories of prints: handtools, food processing, animal management, water management and hydroponics. Such open source hardware is attractive to all types of small farmers as it provides control for farmers over their own equipment; this is exemplified by Open Source Ecology, Farm Hack and FarmBot.

Standards

Standards regulate production methods and in some cases final output for organic agriculture. Standards may be voluntary or legislated. As early as the 1970s private associations certified organic producers. In the 1980s, governments began to produce organic production guidelines. In the 1990s, a trend toward legislated standards began, most notably with the 1991 EU-Eco-regulation developed for European Union, which set standards for 12 countries, and a 1993 UK program. The EU's program was followed by a Japanese program in 2001, and in 2002 the U.S. created the National Organic Program (NOP). As of 2007 over 60 countries regulate organic farming (IFOAM 2007:11). In 2005 IFOAM created the Principles of Organic Agriculture, an international guideline for certification criteria. Typically the agencies accredit certification groups rather than individual farms.

Organic production materials used in and foods are tested independently by the Organic Materials Review Institute.

Composting

Using manure as a fertiliser risks contaminating food with animal gut bacteria, including pathogenic strains of E. coli that have caused fatal poisoning from eating organic food. To combat this risk, USDA organic standards require that manure must be sterilized through high temperature thermophilic composting. If raw animal manure is used, 120 days must pass before the crop is harvested if the final product comes into direct contact with the soil. For products that don't directly contact soil, 90 days must pass prior to harvest.

Economics

The economics of organic farming, a subfield of agricultural economics, encompasses the entire process and effects of organic farming in terms of human society, including social costs, opportunity costs, unintended consequences, information asymmetries, and economies of scale. Although the scope of economics is broad, agricultural economics tends to focus on maximizing yields and efficiency at the farm level. Economics takes an anthropocentric approach to the value of the natural world: biodiversity, for example, is considered beneficial only to the extent that it is valued by people and increases profits. Some entities such as the European Union subsidize organic farming, in large part because these countries want to account for the externalities of reduced water use, reduced water contamination, reduced soil erosion, reduced carbon emissions, increased biodiversity, and assorted other benefits that result from organic farming.

Traditional organic farming is labor and knowledge-intensive whereas conventional farming is capital-intensive, requiring more energy and manufactured inputs.

Organic farmers in California have cited marketing as their greatest obstacle.

Geographic Producer Distribution

The markets for organic products are strongest in North America and Europe, which as of 2001 are estimated to have $6 and $8 billion respectively of the $20 billion global market. As of 2007 Australasia has 39% of the total organic farmland, including Australia's 1,180,000 hectares (2,900,000 acres) but 97 percent of this land is sprawling rangeland (2007:35). US sales are 20x as much. Europe farms 23 percent of global organic farmland (6,900,000 ha (17,000,000 acres)), followed by Latin America with 19 percent (5.8 million hectares - 14.3 million acres). Asia has 9.5 percent while North America has 7.2 percent. Africa has 3 percent.

Besides Australia, the countries with the most organic farmland are Argentina (3.1 million hectares - 7.7 million acres), China (2.3 million hectares - 5.7 million acres), and the United States (1.6 million hectares - 4 million acres). Much of Argentina's organic farmland is pasture, like that of Australia (2007:42). Spain, Germany, Brazil (the world's largest agricultural exporter), Uruguay, and the UK follow the United States in the amount of organic land (2007:26).

In the European Union (EU25) 3.9% of the total utilized agricultural area was used for organic production in 2005. The countries with the highest proportion of organic land were Austria (11%) and Italy (8.4%), followed by the Czech Republic and Greece (both 7.2%). The lowest figures were shown for Malta (0.1%), Poland (0.6%) and Ireland (0.8%). In 2009, the proportion of organic land in the EU grew to 4.7%. The countries with highest share of agricultural land were Liechtenstein (26.9%), Austria (18.5%) and Sweden (12.6%). 16% of all farmers in Austria produced organically in 2010. By the same year the proportion of organic land increased to 20%.: In 2005 168,000 ha (415,000 ac) of land in Poland was under organic management. In 2012, 288,261 hectares (712,308 acres) were under organic production, and there were about 15,500 organic farmers; retail sales of organic products were EUR 80 million in 2011. As of 2012 organic exports were part of the government's economic development strategy.

After the collapse of the Soviet Union in 1991, agricultural inputs that had previously been purchased from Eastern bloc countries were no longer available in Cuba, and many Cuban farms converted to organic methods out of necessity. Consequently, organic agriculture is a mainstream practice in Cuba, while it remains an alternative practice in most other countries. Cuba's organic strategy includes development of genetically modified crops; specifically corn that is resistant to the palomilla moth

Growth

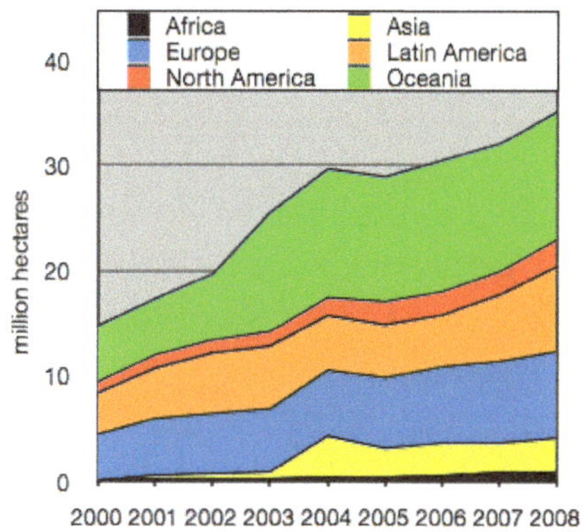

Organic farmland by world region (2000-2008)

In 2001, the global market value of certified organic products was estimated at USD $20 billion. By 2002, this was USD $23 billion and by 2015 more than USD $43 billion. By 2014, retail sales of organic products reached USD $80 billion worldwide. North America and Europe accounted for more than 90% of all organic product sales.

Organic agricultural land increased almost fourfold in 15 years, from 11 million hectares in 1999 to 43.7 million hectares in 2014. Between 2013 and 2014, organic agricultural land grew by 500,000 hectares worldwide, increasing in every region except Latin America. During this time period, Europe's organic farmland increased 260,000 hectares to 11.6 million total (+2.3%), Asia's increased 159,000 hectares to 3.6 million total (+4.7%), Africa's increased 54,000 hectares to 1.3 million total (+4.5%), and North America's increased 35,000 hectares to 3.1 million total (+1.1%). As of 2014, the country with the most organic land was Australia (17.2 million hectares), followed by Argentina (3.1 million hectares), and the United States (2.2 million hectares).

In 2013, the number of organic producers grew by almost 270,000, or more than 13%. By 2014, there were a reported 2.3 million organic producers in the world. Most of the total global increase took place in the Philippines, Peru, China, and Thailand. Overall, the majority of all organic producers are in India (650,000 in 2013), Uganda (190,552 in 2014), Mexico (169,703 in 2013) and the Philippines (165,974 in 2014).

Productivity

Studies comparing yields have had mixed results. These differences among findings can often be attributed to variations between study designs including differences in the crops studied and the methodology by which results were gathered.

A 2012 meta-analysis found that productivity is typically lower for organic farming than conventional farming, but that the size of the difference depends on context and in some cases may be very small. While organic yields can be lower than conventional yields, another meta-analysis published in Sustainable Agriculture Research in 2015, concluded that certain organic on-farm practices could help narrow this gap. Timely weed management and the application of manure in conjunction with legume forages/cover crops were shown to have positive results in increasing organic corn and soybean productivity. More experienced organic farmers were also found to have higher yields than other organic farmers who were just starting out.

Another meta-analysis published in the journal Agricultural Systems in 2011 analyzed 362 datasets and found that organic yields were on average 80% of conventional yields. The author's found that there are relative differences in this yield gap based on crop type with crops like soybeans and rice scoring higher than the 80% average and crops like wheat and potato scoring lower. Across global regions, Asia and Central Europe were found to have relatively higher yields and Northern Europe relatively lower than the average.

A 2007 study compiling research from 293 different comparisons into a single study to assess the overall efficiency of the two agricultural systems has concluded that "organic methods could produce enough food on a global per capita basis to sustain the current human population, and potentially an even larger population, without increasing the agricultural land base." The researchers also found that while in developed countries, organic systems on average produce 92% of the yield produced by conventional agriculture, organic systems produce 80% more than conventional farms in developing countries, because the materials needed for organic farming are more accessible than synthetic farming materials to farmers in some poor countries. This study was strongly contested by another study published in 2008, which stated, and was entitled, "Organic agriculture cannot feed the world" and said that the 2007 came up with "a major overestimation of the productivity of OA" "because data are misinterpreted and calculations accordingly are erroneous." Additional research needs to be conducted in the future to further clarify these claims.

Long Term Studies

A study published in 2005 compared conventional cropping, organic animal-based cropping, and organic legume-based cropping on a test farm at the Rodale Institute over 22 years. The study found that "the crop yields for corn and soybeans were similar in the organic animal, organic legume, and conventional farming systems". It also found that "significantly less fossil energy was expended to produce corn in the Rodale Institute's organic animal and organic legume systems than in the conventional production system. There was little difference in energy input between the different treatments for producing soybeans. In the organic systems, synthetic fertilizers and pesticides were generally not used". As of 2013 the Rodale study was ongoing and a thirty-year anniversary report was published by Rodale in 2012.

A long-term field study comparing organic/conventional agriculture carried out over 21 years in Switzerland concluded that "Crop yields of the organic systems averaged over 21 experimental years at 80% of the conventional ones. The fertilizer input, however, was 34 – 51% lower, indicating an efficient production. The organic farming systems used 20 – 56% less energy to produce a crop unit and per land area this difference was 36 – 53%. In spite of the considerably lower pesticide input the quality of organic products was hardly discernible from conventional analytically and even came off better in food preference trials and picture creating methods"

Profitability

In the United States, organic farming has been shown to be 2.9 to 3.8 times more profitable for the farmer than conventional farming when prevailing price premiums are taken into account. Globally, organic farming is between 22 and 35 percent more profitable for farmers than conventional methods, according to a 2015 meta-analysis of studies conducted across five continents.

The profitability of organic agriculture can be attributed to a number of factors. First, organic farmers do not rely on synthetic fertilizer and pesticide inputs, which can be costly. In addition, organic foods currently enjoy a price premium over conventionally produced foods, meaning that organic farmers can often get more for their yield.

The price premium for organic food is an important factor in the economic viability of organic farming. In 2013 there was a 100% price premium on organic vegetables and a 57% price premium for organic fruits. These percentages are based on wholesale fruit and vegetable prices, available through the United States Department of Agriculture's Economic Research Service. Price premiums exist not only for organic versus nonorganic crops, but may also vary depending on the venue where the product is sold: farmers markets, grocery stores, or wholesale to restaurants. For many producers, direct sales at farmers markets are most profitable because the farmer receives the entire markup, however this is also the most time and labor-intensive approach.

There have been signs of organic price premiums narrowing in recent years, which lowers the economic incentive for farmers to convert to or maintain organic production methods. Data from 22 years of experiments at the Rodale Institute found that, based on the current yields and production costs associated with organic farming in the United States, a price premium of only 10% is required to achieve parity with conventional farming. A separate study found that on a global scale, price premiums of only 5-7% percent were needed to break even with conventional methods. Without the price premium, profitability for farmers is mixed.

For markets and supermarkets organic food is profitable as well, and is generally sold at significantly higher prices than non-organic food.

Energy Efficiency

In the most recent assessments of the energy efficiency of organic versus conventional agriculture, results have been mixed regarding which form is more carbon efficient. Organic farm systems have more often than not been found to be more energy efficient, however, this is not always the case. More than anything, results tend to depend upon crop type and farm size.

A comprehensive comparison of energy efficiency in grain production, produce yield, and animal husbandry concluded that organic farming had a higher yield per unit of energy over the vast majority of the crops and livestock systems. For example, two studies - both comparing organically- versus conventionally-farmed apples - declare contradicting results, one saying organic farming is more energy efficient, the other saying conventionally is more efficient.

It has generally been found that the labor input per unit of yield was higher for organic systems compared with conventional production.

Sales and Marketing

Most sales are concentrated in developed nations. In 2008, 69% of Americans claimed to occasionally buy organic products, down from 73% in 2005. One theory for this change was that consumers were substituting "local" produce for "organic" produce.

Distributors

The USDA requires that distributors, manufacturers, and processors of organic products be certified by an accredited state or private agency. In 2007, there were 3,225 certified organic handlers, up from 2,790 in 2004.

Organic handlers are often small firms; 48% reported sales below $1 million annually, and 22% between $1 and $5 million per year. Smaller handlers are more likely to sell to independent natural grocery stores and natural product chains whereas large distributors more often market to natural product chains and conventional supermarkets, with a small group marketing to independent natural product stores. Some handlers work with conventional farmers to convert their land to organic with the knowledge that the farmer will have a secure sales outlet. This lowers the risk for the handler as well as the farmer. In 2004, 31% of handlers provided technical support on organic standards or production to their suppliers and 34% encouraged their suppliers to transition to organic. Smaller farms often join together in cooperatives to market their goods more effectively.

93% of organic sales are through conventional and natural food supermarkets and chains, while the remaining 7% of U.S. organic food sales occur through farmers' markets, foodservices, and other marketing channels.

Direct-to-consumer Sales

In the 2012 Census, direct-to-consumer sales equaled $1.3 billion, up from $812 million in 2002, an increase of 60 percent. The number of farms that utilize direct-to-consumer sales was 144,530 in 2012 in comparison to 116,733 in 2002. Direct-to-consumer sales include farmers markets, community supported agriculture (CSA), on-farm stores, and roadside farm stands. Some organic farms also sell products direct to retailer, direct to restaurant and direct to institution. According to the 2008 Organic Production Survey, approximately 7% of organic farm sales went direct-to-consumers, 10% went direct to retailers, and approximately 83% went into wholesale markets. In comparison, only 0.4% of the value of convention agricultural commodities went direct-to-consumers.

While not all products sold at farmer's markets are certified organic, this direct-to-consumer avenue has become increasingly popular in local food distribution and has grown substantially since 1994. In 2014, there were 8,284 farmer's markets in comparison to 3,706 in 2004 and 1,755 in 1994, most of which are found in populated areas such as the Northeast, Midwest, and West Coast.

Labor and Employment

Organic production is more labor-intensive than conventional production. On the one hand, this increased labor cost is one factor that makes organic food more expensive. On the other hand, the increased need for labor may be seen as an "employment dividend" of organic farming, providing more jobs per unit area than conventional systems. The 2011 UNEP Green Economy Report suggests that "[a]n increase in investment in green agriculture is projected to lead to growth in employment of about 60 per cent compared with current levels" and that "green agriculture investments could create 47 million additional jobs compared with BAU2 over the next 40 years." The UNEP also argues that "[b]y greening agriculture and food distribution, more calories per person per day, more jobs and business opportunities especially in rural areas, and market-access opportunities, especially for developing countries, will be available."

World's Food Security

In 2007 the United Nations Food and Agriculture Organization (FAO) said that organic agriculture often leads to higher prices and hence a better income for farmers, so it should be promoted. However, FAO stressed that by organic farming one could not feed the current mankind, even less the bigger future population. Both data and models showed then that organic farming was far from sufficient. Therefore, chemical fertilizers were needed to avoid hunger. Other analysis by many agribusiness executives, agricultural and ecological scientists, and international agriculture experts revealed the opinion that organic farming would not only increase the world's food supply, but might be the only way to eradicate hunger.

FAO stressed that fertilizers and other chemical inputs can much increase the production, particularly in Africa where fertilizers are currently used 90% less than in Asia. For example, in Malawi the yield has been boosted using seeds and fertilizers. FAO also calls for using biotechnology, as it can help smallholder farmers to improve their income and food security.

Also NEPAD, development organization of African governments, announced that feeding Africans and preventing malnutrition requires fertilizers and enhanced seeds.

According to a more recent study in ScienceDigest, organic best management practices shows an average yield only 13% less than conventional. In the world's poorer nations where most of the world's hungry live, and where conventional agriculture's expensive inputs are not affordable by the majority of farmers, adopting organic management actually increases yields 93% on average, and could be an important part of increased food security.

Capacity Building in Developing Countries

Organic agriculture can contribute to ecologically sustainable, socio-economic development, especially in poorer countries. The application of organic principles enables employment of local resources (e.g., local seed varieties, manure, etc.) and therefore cost-effectiveness. Local and international markets for organic products show tremendous growth prospects and offer creative producers and exporters excellent opportunities to improve their income and living conditions.

Organic agriculture is knowledge intensive. Globally, capacity building efforts are underway, including localized training material, to limited effect. As of 2007, the International Federation of Organic Agriculture Movements hosted more than 170 free manuals and 75 training opportunities online.

In 2008 the United Nations Environmental Programme (UNEP) and the United Nations Conference on Trade and Development (UNCTAD) stated that "organic agriculture can be more conducive to food security in Africa than most conventional production systems, and that it is more likely to be sustainable in the long-term" and that "yields had more than doubled where organic, or near-organic practices had been used" and that soil fertility and drought resistance improved.

Millennium Development Goals

The value of organic agriculture (OA) in the achievement of the Millennium Development Goals (MDG), particularly in poverty reduction efforts in the face of climate change, is shown by its contribution to both income and non-income aspects of the MDGs. These benefits are expected to continue in the post-MDG era. A series of case studies conducted in selected areas in Asian countries by the Asian Development Bank Institute (ADBI) and published as a book compilation by ADB in Manila document these contributions to both income and non-income aspects of the MDGs. These include poverty alleviation by way of higher incomes, improved farmers' health owing to less chemical exposure, integration of sustainable principles into rural development policies, improvement of access to safe water and sanitation, and expansion of global partnership for development as small farmers are integrated in value chains.

A related ADBI study also sheds on the costs of OA programs and set them in the context of the costs of attaining the MDGs. The results show considerable variation across the case studies, suggesting that there is no clear structure to the costs of adopting OA. Costs depend on the efficiency of the OA adoption programs. The lowest cost programs were more than ten times less expensive than the highest cost ones. However, further analysis of the gains resulting from OA adoption reveals that the costs per person taken out of poverty was much lower than the estimates of the World Bank, based on income growth in general or based on the detailed costs of meeting some of the more quantifiable MDGs (e.g., education, health, and environment).

Externalities

Agriculture imposes negative externalities (uncompensated costs) upon society through public land and other public resource use, biodiversity loss, erosion, pesticides, nutrient runoff, subsidized water usage, subsidy payments and assorted other problems. Positive externalities include self-reliance, entrepreneurship, respect for nature, and air quality. Organic methods reduce some of these costs. In 2000 uncompensated costs for 1996 reached 2,343 million British pounds or £208 per ha (£84.20/ac). A study of practices in the USA published in 2005 concluded that cropland costs the economy approximately 5 to 16 billion dollars ($30–96/ha - $12–39/ac), while livestock production costs 714 million dollars. Both studies recommended reducing externalities. The 2000 review included reported pesticide poisonings but did not include speculative chronic health effects of pesticides, and the 2004 review relied on a 1992 estimate of the total impact of pesticides.

It has been proposed that organic agriculture can reduce the level of some negative externalities from (conventional) agriculture. Whether the benefits are private or public depends upon the division of property rights.

Several surveys and studies have attempted to examine and compare conventional and organic systems of farming and have found that organic techniques, while not without harm, are less damaging than conventional ones because they reduce levels of biodiversity less than conventional systems do and use less energy and produce less waste when calculated per unit area.

A 2003 to 2005 investigation by the Cranfield University for the Department for Environment Food and Rural Affairs in the UK found that it is difficult to compare the Global Warming Potential (GWP), acidification and eutrophication emissions but "Organic production often results in increased burdens, from factors such as N leaching and N2O emissions", even though primary energy use was less for most organic products. N_2O is always the largest GWP contributor except in tomatoes. However, "organic tomatoes always incur more burdens (except pesticide use)". Some emissions were lower "per area", but organic farming always required 65 to 200% more field area than non-organic farming. The numbers were highest for bread wheat (200+ % more) and potatoes (160% more).

The situation was shown dramatically in a comparison of a modern dairy farm in Wisconsin with one in New Zealand in which the animals grazed extensively. Using total farm emissions per kg milk produced as a parameter, the researchers showed that production of methane from belching was higher in the New Zealand farm, while carbon dioxide production was higher in the Wisconsin farm. Output of nitrous oxide, a gas with an estimated global warming potential 310 times that of carbon dioxide was also higher in the New Zealand farm. Methane from manure handling was similar in the two types of farm. The explanation for the finding relates to the different diets used on these farms, being based more completely on forage (and hence more fibrous) in New Zealand and containing less concentrate than in Wisconsin. Fibrous diets promote a higher proportion of acetate in the gut of ruminant animals, resulting in a higher production of methane that must be released by belching. When cattle are given a diet containing some concentrates (such as corn and soybean meal) in addition to grass and silage, the pattern of ruminal fermentation alters from acetate to mainly propionate. As a result, methane production is reduced. Capper et al. compared the environmental impact of US dairy production in 1944 and 2007. They calculated that the carbon "footprint" per billion kg (2.2 billion lb) of milk produced in 2007 was 37 percent that of equivalent milk production in 1944.

Environmental Impact and Emissions

Researchers at Oxford university analyzed 71 peer-reviewed studies and observed that organic products are sometimes worse for the environment. Organic milk, cereals, and pork generated higher greenhouse gas emissions per product than conventional ones but organic beef and olives had lower emissions in most studies. Usually organic products required less energy, but more land. Per unit of product, organic produce generates higher nitrogen leaching, nitrous oxide emissions, ammonia emissions, eutrophication and acidification potential than when conventionally grown. Other differences were not significant. The researchers concluded, as there is not singular way of doing conventional or organic farming, that the debate should go beyond the conventional vs organic debate, and more about finding specific solutions to specific circumstances.

Proponents of organic farming have claimed that organic agriculture emphasizes closed nutrient cycles, biodiversity, and effective soil management providing the capacity to mitigate and even reverse the effects of climate change and that organic agriculture can decrease fossil fuel emissions. "The carbon sequestration efficiency of organic systems in temperate climates is almost double (575-700 kg carbon per ha per year - 510-625 lb/ac/an) that of conventional treatment of soils, mainly owing to the use of grass clovers for feed and of cover crops in organic rotations."

Critics of organic farming methods believe that the increased land needed to farm organic food could potentially destroy the rainforests and wipe out many ecosystems.

Nutrient Leaching

According to the meta-analysis of 71 studies, nitrogen leaching, nitrous oxide emissions, ammonia emissions, eutrophication potential and acidification potential were higher for organic products, although in one study "nitrate leaching was 4.4-5.6 times higher in conventional plots than organic plots".

Excess nutrients in lakes, rivers, and groundwater can cause algal blooms, eutrophication, and subsequent dead zones. In addition, nitrates are harmful to aquatic organisms by themselves.

Land use

The Oxford meta-analysis of 71 studies proved that organic farming requires 84% more land, mainly due to lack of nutrients but sometimes due to weeds, diseases or pests, lower yielding animals and land required for fertility building crops. While organic farming does not necessarily save land for wildlife habitats and forestry in all cases, the most modern breakthroughs in organic are addressing these issues with success.

Professor Wolfgang Branscheid says that organic animal production is not good for the environment, because organic chicken requires doubly as much land as "conventional" chicken and organic pork a quarter more. According to a calculation by Hudson Institute, organic beef requires triply as much land. On the other hand, certain organic methods of animal husbandry have been shown to restore desertified, marginal, and/ or otherwise unavailable land to agricultural productivity and wildlife. Or by getting both forage and cash crop production from the same fields simultaneously, reduce net land use.

In England organic farming yields 55% of normal yields. While in other regions of the world, organic methods have started producing record yields.

Pesticides

A sign outside of an organic apple orchard in Pateros, Washington reminding orchardists not to spray pesticides on these trees.

In organic farming synthetic pesticides are generally prohibited. A chemical is said to be synthetic if it does not already exist in the natural world. But the organic label goes further and usually prohibit compounds that exist in nature if they are produced by Chemical synthesis. So the prohibition is also about the method of production and not only the nature of the compound.

An non exhaustive list of organic approved pesticides with theirs Median lethal dose

- Copper(II) sulfate is used as a fungicide and is also used in conventional agriculture (LD_{50} 300 mg/kg). Conventional agriculture has the option to use the less toxic Mancozeb (LD_{50} 4,500 to 11,200 mg/kg)
- Boric acid is used as stomach poison that target insects (LD_{50}: 2660 mg/kg).
- Pyrethrin comes from chemicals extracted from flowers of the genus Pyrethrum (LD_{50} of 370 mg/kg). Its potent toxicity is used to control insects.
- Lime sulphur (aka calcium polysulfide) and sulfur are considered to be allowed, synthetic

materials (LD_{50}: 820 mg/kg)

- Rotenone is a powerful insecticide that was used to control insects (LD_{50}: 132 mg/kg). Despite the high toxicity of Rotenone to aquatic life and some links to Parkinson disease the compound is still allowed in organic farming as it is a naturally occurring compound.
- Bromomethane is a gas that is still used in the nurseries of Strawberry organic farming

- Azadirachtin is a wide spectrum very potent insecticide. Almost non toxic to mammals (LD_{50} in rats is > 3,540 mg/kg) but affects beneficial insects.

Food Quality and Safety

While there may be some differences in the amounts of nutrients and anti-nutrients when organically produced food and conventionally produced food are compared, the variable nature of food production and handling makes it difficult to generalize results, and there is insufficient evidence to make claims that organic food is safer or healthier than conventional food. Claims that organic food tastes better are not supported by evidence.

Soil Conservation

Supporters claim that organically managed soil has a higher quality and higher water retention. This may help increase yields for organic farms in drought years. Organic farming can build up soil organic matter better than conventional no-till farming, which suggests long-term yield benefits from organic farming. An 18-year study of organic methods on nutrient-depleted soil concluded that conventional methods were superior for soil fertility and yield for nutrient-depleted soils in cold-temperate climates, arguing that much of the benefit from organic farming derives from imported materials that could not be regarded as self-sustaining.

In *Dirt: The Erosion of Civilizations*, geomorphologist David Montgomery outlines a coming crisis from soil erosion. Agriculture relies on roughly one meter of topsoil, and that is being depleted ten times faster than it is being replaced. No-till farming, which some claim depends upon pesticides, is one way to minimize erosion. However, a 2007 study by the USDA's Agricultural Research Service has found that manure applications in tilled organic farming are better at building up the soil than no-till.

Biodiversity

The conservation of natural resources and biodiversity is a core principle of organic production. Three broad management practices (prohibition/reduced use of chemical pesticides and inorganic fertilizers; sympathetic management of non-cropped habitats; and preservation of mixed farming) that are largely intrinsic (but not exclusive) to organic farming are particularly beneficial for farmland wildlife. Using practices that attract or introduce beneficial insects, provide habitat for birds and mammals, and provide conditions that increase soil biotic diversity serve to supply vital ecological services to organic production systems. Advantages to certified organic operations that implement these types of production practices include: 1) decreased dependence on outside fertility inputs; 2) reduced pest management costs; 3) more reliable sources of clean water; and 4) better pollination.

Nearly all non-crop, naturally occurring species observed in comparative farm land practice studies show a preference for organic farming both by abundance and diversity. An average of 30% more species inhabit organic farms. Birds, butterflies, soil microbes, beetles, earthworms, spiders, vegetation, and mammals are particularly affected. Lack of herbicides and pesticides improve biodiversity fitness and population density. Many weed species attract beneficial insects that improve soil qualities and forage on weed pests. Soil-bound organisms often benefit because of increased bacteria populations due to natural fertilizer such as manure, while experiencing reduced intake of herbicides and pesticides. Increased biodiversity, especially from beneficial soil microbes and mycorrhizae have been proposed as an explanation for the high yields experienced by some organic plots, especially in light of the differences seen in a 21-year comparison of organic and control fields.

Biodiversity from organic farming provides capital to humans. Species found in organic farms enhance sustainability by reducing human input (e.g., fertilizers, pesticides).

The USDA's Agricultural Marketing Service (AMS) published a *Federal Register* notice on 15 January 2016, announcing the National Organic Program (NOP) final guidance on Natural Resources and Biodiversity Conservation for Certified Organic Operations. Given the broad scope of natural resources which includes soil, water, wetland, woodland and wildlife, the guidance provides examples of practices that support the underlying conservation principles and demonstrate compliance with USDA organic regulations § 205.200. The final guidance provides organic certifiers and farms with examples of production practices that support conservation principles and comply with the USDA organic regulations, which require operations to maintain or improve natural resources. The final guidance also clarifies the role of certified operations (to submit an OSP to a certifier), certifiers (ensure that the OSP describes or lists practices that explain the operator's monitoring plan and practices to support natural resources and biodiversity conservation), and inspectors (onsite inspection) in the implementation and verification of these production practices.

A wide range of organisms benefit from organic farming, but it is unclear whether organic methods confer greater benefits than conventional integrated agri-environmental programs. Organic farming is often presented as a more biodiversity-friendly practice, but the generality of the beneficial effects of organic farming is debated as the effects appear often species- and context-dependent, and current research has highlighted the need to quantify the relative effects of local- and landscape-scale management on farmland biodiversity. There are four key issues when comparing the impacts on biodiversity of organic and conventional farming: (1) It remains unclear whether a holistic whole-farm approach (i.e. organic) provides greater benefits to biodiversity than carefully targeted prescriptions applied to relatively small areas of cropped and/or non-cropped habitats within conventional agriculture (i.e. agri-environment schemes); (2) Many comparative studies encounter methodological problems, limiting their ability to draw quantitative conclusions; (3) Our knowledge of the impacts of organic farming in pastoral and upland agriculture is limited; (4) There remains a pressing need for longitudinal, system-level studies in order to address these issues and to fill in the gaps in our knowledge of the impacts of organic farming, before a full appraisal of its potential role in biodiversity conservation in agroecosystems can be made.

References

- Coleman, Eliot (1995), The New Organic Grower: A Master's Manual of Tools and Techniques for the Home and Market Gardener (2nd ed.), pp. 65, 108, ISBN 978-0930031756.
- Horne, Paul Anthony (2008). Integrated pest management for crops and pastures. CSIRO Publishing. p. 2. ISBN 978-0-643-09257-0.
- Vogt G (2007). Lockeretz W, ed. Chapter 1: The Origins of Organic Farming. Organic Farming: An International History. CABI Publishing. pp. 9–30. ISBN 9780851998336.
- Pamela Ronald; Raoul Admachak (April 2008). "Tomorrow's Table: Organic Farming, Genetics and the Future of Food". Oxford University Press. ISBN 0195301757.
- Halberg, Niels (2006). Global development of organic agriculture: challenges and prospects. CABI. p. 297. ISBN 978-1-84593-078-3.
- Blair, Robert. (2012). Organic Production and Food Quality: A Down to Earth Analysis. Wiley-Blackwell, Oxford, UK. ISBN 978-0-8138-1217-5
- "USDSA Guidance Natural Resources and Biodiversity Conservation" (PDF). Agricultural Marketing Service, National Organic Program. United States Department of Agriculture. 15 January 2016. Retrieved 5 March 2016.
- Henckel, Laura (20 May 2015). "Organic fields sustain weed metacommunity dynamics in farmland landscapes". Proceedings of the Royal Society B. Retrieved 28 February 2016.
- "USDA List of Allowed and Prohibited Substances in Organic Agriculture". USDA List of Allowed and Prohibited Substances in Organic Agriculture. USDA. 4 April 2016. Retrieved 6 April 2016.
- Arsenault, Chris. "Only 60 Years of Farming Left If Soil Degradation Continues". Scientific American. Retrieved 29 May 2016.
- "Can organic food feed the world? New study sheds light on debate over organic vs. conventional agriculture". Science Daily. Retrieved 2 March 2014.
- De Schutter, Olivier. "Report submitted by the Special Rapporteur on the right to food" (PDF). United Nations. Retrieved 3 March 2014.
- Undersander, Dan; et al. "Pastures for Profit: A Guide to Rotational Grazing" (PDF). University of Wisconsin. Cooperative extension publishing. Retrieved 24 February 2014.

6

Evolution of Plant Breeding

Plant breeding started with the domestication of the first agricultural plants, almost 9000 to 11000 years back. Laws of inheritance formed the basis of the new science of genetics, causing many plant scientists to also research and to contribute to improving crop production through plant breeding. The evolution of plant breeding explained in this chapter is of great importance to broaden the existing knowledge on plant breeding.

Plant breeding started with sedentary agriculture, particularly the domestication of the first agricultural plants, a practice which is estimated to date back 9,000 to 11,000 years. Initially, early human farmers selected food plants with particular desirable characteristics and used these as a seed source for subsequent generations, resulting in an accumulation of characteristics over time. In time however, experiments began with deliberate hybridization, the science and understanding of which was greatly enhanced by the work of Gregor Mendel. Mendel's work ultimately led to the new science of genetics. Modern plant breeding is applied genetics, but its scientific basis is broader, covering molecular biology, cytology, systematics, physiology, pathology, entomology, chemistry, and statistics (biometrics). It has also developed its own technology. Plant breeding efforts are divided into a number of different historical landmarks.

Early Plant Breeding

Domestication

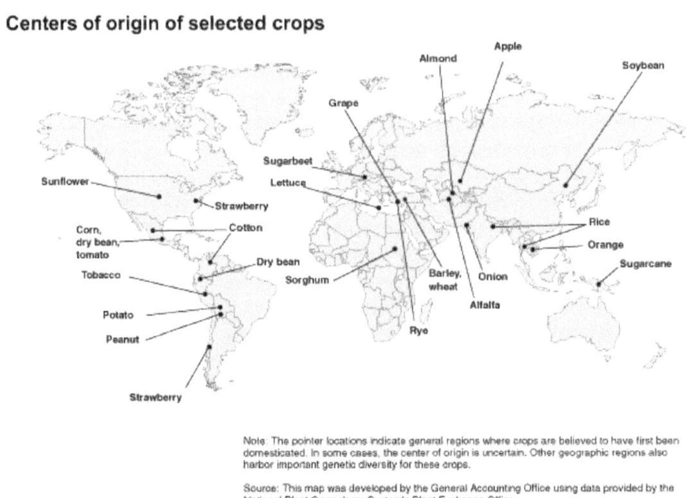

This map shows the sites of domestication for a number of crops. Places where crops were initially domesticated are called centers of origin

Domestication of plants is an artificial selection process conducted by humans to produce plants that have more desirable traits than wild plants, and which renders them dependent on artificial usually enhanced environments for their continued existence. The practice is estimated to date back 9,000-11,000 years. Many crops in present-day cultivation are the result of domestication in ancient times, about 5,000 years ago in the Old World and 3,000 years ago in the New World. In the Neolithic period, domestication took a minimum of 1,000 years and a maximum of 7,000 years. Today, all principal food crops come from domesticated varieties. Almost all the domesticated plants used today for food and agriculture were domesticated in the centers of origin. In these centers there is still a great diversity of closely related wild plants, so-called crop wild relatives, that can also be used for improving modern cultivars by plant breeding.

A plant whose origin or selection is due primarily to intentional human activity is called a cultigen, and a cultivated crop species that has evolved from wild populations due to selective pressures from traditional farmers is called a landrace. Landraces, which can be the result of natural forces or domestication, are plants or animals that are ideally suited to a particular region or environment. An example are the landraces of rice, *Oryza sativa* subspecies *indica*, which was developed in South Asia, and *Oryza sativa* subspecies *japonica*, which was developed in China.

For more on the mechanisms of domestication.

Columbian Exchange

Humans have traded useful plants from distant lands for centuries, and plant hunters have been sent to bring plants back for cultivation. Human agriculture has had two important results: the plants most favoured by humans came to be grown in many places and (2) gardens and farms have provided some opportunities for plants to interbreed that would not have been possible for their wild ancestors. Columbus's arrival in America in 1492 triggered unprecedented transfer of plant resources between Europe and the New World.

Scientific Plant Breeding

Gregor Mendel's experiments with plant hybridization led to his laws of inheritance. This work became well known in the 1900s and formed the basis of the new science of genetics, which stimulated research by many plant scientists dedicated to improving crop production through plant breeding.

However, successful commercial plant breeding concerns began to be founded from the late 19th century. Gartons Agricultural Plant Breeders in England was established in the 1890s by John Garton, who was one of the first to cross-pollinate agricultural plants and commercialize the newly created varieties. He began experimenting with the artificial cross pollination firstly of cereal plants, then herbage species and root crops and developed far reaching techniques in plant breeding.

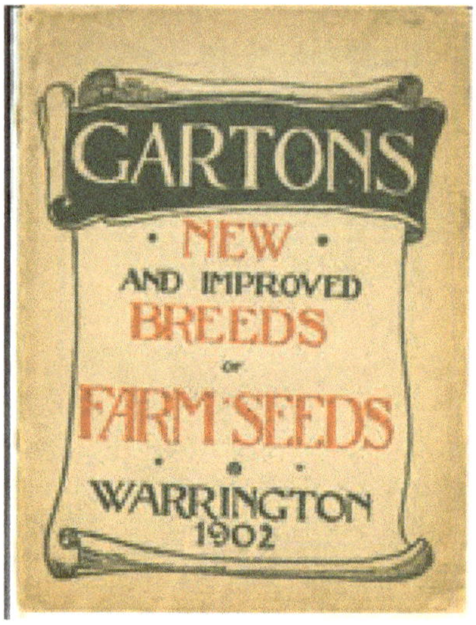

Garton's catalogue from 1902

From 1904 to World War II in Italy, Nazareno Strampelli created a number of wheat hybrids. His work allowed Italy to increase crop production during the so-called "Battle for Grain" (1925–1940) and some varieties were exported to foreign countries, such as Argentina, Mexico, and China. Strampelli's work laid the foundations for Norman Borlaug and the Green Revolution.

Green Revolution

In 1908, George Harrison Shull described heterosis, also known as hybrid vigor. Heterosis describes the tendency of the progeny of a specific cross to outperform both parents. The detection of the usefulness of heterosis for plant breeding has led to the development of inbred lines that reveal a heterotic yield advantage when they are crossed. Maize was the first species where heterosis was widely used to produce hybrids.

By the 1920s, statistical methods were developed to analyze gene action and distinguish heritable variation from variation caused by environment. In 1933 another important breeding technique, cytoplasmic male sterility (CMS), developed in maize, was described by Marcus Morton Rhoades. CMS is a maternally inherited trait that makes the plant produce sterile pollen. This enables the production of hybrids without the need for labor-intensive detasseling.

These early breeding techniques resulted in large yield increase in the United States in the early 20th century. Similar yield increases were not produced elsewhere until after World War II, the Green Revolution increased crop production in the developing world in the 1960s. This remarkable improvement was based on three essential crops. First came the development of hybrid maize, then high-yielding and input-responsive "semi-dwarf wheat" (for which the CIMMYT breeder N.E. Borlaug received the Nobel prize for peace in 1970), and third came high-yielding "short statured rice" cultivars. Similarly notable improvements were achieved in other crops like sorghum and alfalfa.

Molecular Genetics and Bio-revolution

Intensive research in molecular genetics has led to the development of recombinant DNA technology (popularly called genetic engineering). Advancement in biotechnological techniques has opened many possibilities for breeding crops. Thus, while mendelian genetics allowed plant breeders to perform genetic transformations in a few crops, molecular genetics has provided the key to both the manipulation of the internal genetic structure, and the "crafting" of new cultivars according to a pre-determined plan.

7

Acts Related to Plant Breeding

This chapter focuses on the acts affiliated to plant breeding. This chapter lists the following acts like the protection of plant varieties and farmers rights act, 2001, plant variety protection act of 1970, plant patent act of 1930 and international union for the protection of new varieties of plants. This chapter will help the readers in understanding these acts.

Protection of Plant Varieties and Farmers' Rights Act, 2001

The Protection of Plant Variety and Farmers Right Act, 2001 (PPVFR Act) is an Act of the Parliament of India enacted to provide for the establishment of an effective system for protection of plant varieties, the rights of farmers and plant breeders, and to encourage the development and cultivation of new varieties of plants. This act received the assent of the President of India on the October 30, 2001.

The PPV&FR Act, 2001 was enacted to grant intellectual property rights to plant breeders, researchers and farmers who have developed any new or extant plant varieties. The Intellectual Property Right granted under PPV&FR Act, 2001 is a dual right- one is for the variety and the other is for the denomination assigned to it by the breeder. The rights granted under this Act are heritable and assignable and only registration of a plant variety confers the right. Essentially Derived Varieties (EDV) can also be registered under this Act and it may be new or extant. Farmers are entitled to save, use, sow, re-sow, exchange or sell their farm produce including seed of a registered variety in an unbranded manner. Farmers' varieties are eligible for registration and farmers are totally exempted from payment of any fee in any proceedings under this Act. The period of protection for field crops is 15 years and for trees and vines is 18 years and for notified varieties it is 15 years from the date of notification under section 5 of Seeds Act, 1966. Annual fee has to be paid every year for maintaining the registration and renewal fee has to be paid for the extended period of registration. Farmers' can claim for compensation if the registered variety fails to provide expected performance under given conditions. The rights granted under this Act are exclusive right to produce, sell, market, distribute, import and export the variety. Civil and criminal remedies are provided for enforcement of breeders' rights and provisions relating to benefit sharing and compulsory licence in case registered variety is not made available to the public at reasonable price are provided. Compensation is also provided for village or rural communities if any registered variety has been developed using any variety in whose evolution such village or local community has contributed significantly. The procedural details and modes of implementing this Act are provided in PPV&FR Rules, 2003. In the present article, we are discussing some of the important legal provisions of this Act alongwith some of the case studies.

Definition

According to sec 2(c), "breeder" means a person or group of persons or a farmer or group of farmers or any institution which has "bred, evolved or developed any variety."

According to sec 2(k), "farmers" means any person who -

- "Cultivates crops by cultivating the land himself; or"
- "Cultivates crops by directly supervising the cultivation or land through any other person; or conserves and preserves, severally or jointly, with any other person any wild species or traditional varieties"; or
- "Adds value to such wild species or traditional varieties through selection and identification of their useful properties."

Plant Genome Saviour Awards

Community of farmers who have engaged in conservation, improvement and preservation of genetic resources of economic plants and their wild relatives particularly in areas identified as agro-biodiversity hotspots are awarded annually from Gene Fund. The name of the award is Plant Genome Saviour Community Award and the amount is INR 10,00,000 for each community. A maximum of five awards are conferred in a year.

The Authority also confers Plant Genome Savior "Farmer Reward" and "Farmer Recognition" to the farmers engaged in the conservation of genetic resources of landraces and wild relatives of economic plants and their improvement through selection and preservation and the material so selected and preserved has been used as donors of gene in varieties registerable under the PPV&FR Act, 2001 (53 of 2001) . There shall be maximum of 10 rewards (citation, memento and cash of INR to be decided by Central Government) and 20 recognitions (consisting of a citation and memento and cash award to be decided by Central Government) in a year.

Plant Variety Protection Act of 1970

The Plant Variety Protection Act of 1970 (PVPA), 7 U.S.C. §§ 2321-2582, is an intellectual property statute in the United States. The PVPA gives breeders up to 25 years of exclusive control over new, distinct, uniform, and stable sexually reproduced or tuber propagated plant varieties. A major expression of plant breeders' rights in the United States, the PVPA grants protection similar to that available through patents, but these legal schemes differ in critical respects. The PVPA should not be confused with plant patents, which are limited to asexually reproduced plants (not including tuber propagated plants).

Basic Provisions

The PVPA confers a limited period of legal control to breeders of sexually reproduced or tuber propagated plant varieties. In order to be eligible for a certificate under the PVPA, a plant variety must satisfy four requirements. First, it must be *new*, in the sense that propagating or harvested material has not been sold or otherwise disposed of for purposes of exploitation for more than one year in the United States, or more than four years in any foreign jurisdiction (or six years in the case of a tree or vine). Second, the variety must be *distinct*—that is, clearly distinguishable from any other publicly known variety. Distinctness may be based on one or more identifiable morphological, physiological, or other characteristics, including commercially valuable characteristics affecting activities such as milling and baking (in the case of wheat). Third, the variety must be *uniform*, in the sense that any variations are describable, predictable, and commercially acceptable. Finally, the variety must be *stable*, in the sense that the variety, when reproduced, will remain unchanged with regard to its essential and distinctive characteristics within a reasonable degree of commercial reliability.

A plant variety certificate gives the breeder the right to exclude others from selling the variety, or offering it for sale, or reproducing it, or importing it, or exporting it, or using it in producing (as distinguished from developing) a hybrid or different variety. The term of protection runs 20 years from the certificate's date of issue, or 25 years in the case of a tree or vine.

Exemptions

Unlike the Patent Act of 1952, the PVPA contains three exemptions that significantly limit the scope of the plant breeder's exclusive right. First, the PVPA's provision safeguarding the "public interest in wide usage" allows the United States Department of Agriculture to declare an otherwise protected variety open on the basis of equitable remuneration to the owner, upon a finding that no more than two years of compulsory licensing of a protected variety is necessary in order to insure an adequate supply of fiber, food, or feed and that the owner is unwilling or unable to meet public demand at a price which may reasonably be deemed fair. Second, the PVPA's "research exemption" declares that the use and reproduction of a protected variety for plant breeding or other bona fide research shall not constitute infringement.

The PVPA's third exemption permits a farmer to save seed from protected varieties and to use such saved seed in the production of a crop without infringement. Prior to 1994, this exemption also allowed farmers to sell such saved seed to others without infringement. Asgrow Seed Company sued Denny and Becky Winterboer over the scope of this exemption. Asgrow won at District Court, and then lost at the Court of Appeals. The scope of the exemption to sell seeds was confirmed and defined by the 1995 Supreme Court decision in *Asgrow Seed Co. v. Winterboer*, 513 U.S. 179 (1995). In 1994, legislation to bring the PVPA into compliance with the 1991 Act of the UPOV Convention also included amendments that eliminated the exemption for sales, but continued to allow farmers to save and replant seed on their own farms without infringement.

Contrasting Plant Variety Certificates, Plant Patents, and Utility Patents

Plant variety certificates should not be confused with plant patents. The Plant Variety Protection Act (PVPA) and the Plant Patent Act of 1930 provide protection based on different types of plant reproductive strategies. The PVPA, which authorizes plant variety protection certificates, protects sexually reproduced and tuber propagated plants, whereas the Plant Patent Act, which authorizes plant patents, is limited to asexually reproduced plants, excluding tuber propagated plants. However, the same plant may be protected under both the Plant Variety Protection Act and the Plant Patent Act of 1930. The most appropriate means of protection may be decided by the applicant based on the applicant's business requirements.

A plant may also be eligible for protection under the Patent Act of 1952, as long as patentability requirements are met. Additionally, the same plant may be protected by a utility patent as well as a plant variety protection certificate and/or a plant patent. The landmark Supreme Court decision of *Diamond v. Chakrabarty*, 447 U.S. 303 (1980), suggested the possibility of securing utility patents on plants previously thought eligible solely for protection under the PVPA. Two decades later, in the 2001 decision of *J.E.M. Ag Supply, Inc. v. Pioneer Hi-Bred International, Inc.*, 534 U.S. 124 (2001), the Court conclusively held that sexually reproduced plants eligible for protection under the PVPA are also eligible for utility patents.

The PVPA and International Law

The Plant Variety Protection Act represents the United States' effort to comply with the *Union pour la Protection des Obtentions Végétales* (also known as UPOV or the International Union for the Protection of New Varieties of Plants), an international treaty concerning plant breeders' rights. The PVPA likewise constitutes part of the United States' compliance with the Trade Related Aspects of Intellectual Property Rights annex of the World Trade Organization treaty, also known as TRIPs.

Plant Patent Act of 1930

The Plant Patent Act of 1930 (enacted on 1930-06-17 as Title III of the Smoot–Hawley Tariff, ch. 497, 46 Stat. 703, codified as 35 U.S.C. Ch. 15) is a United States federal law spurred by the work of Luther Burbank.

This piece of legislation made it possible to patent new varieties of plants, excluding sexual and tuber-propagated plants Plant Variety Protection Act of 1970. In sup-porting the legislation, Thomas Edison testified before Congress in support of the leg-islation and said,

> *"This [bill] will, I feel sure, give us many Burbanks."*

Plant patents PP12, PP13, PP14, PP15, PP16, PP18, PP41, PP65, PP66, PP235, PP266, PP267, PP269, PP290, PP291 and PP1041 were issued to Burbank posthumously.

International Union for the Protection of New Varieties of Plants

The International Union for the Protection of New Varieties of Plants or UPOV (French: *Union internationale pour la protection des obtentions végétales*) is an intergovernmental organization with headquarters in Geneva, Switzerland. The current Secretary-General of UPOV is Francis Gurry.

UPOV was established by the International Convention for the Protection of New Varieties of Plants. The Convention was adopted in Paris in 1961 and revised in 1972, 1978 and 1991. The objective of the Convention is the protection of new varieties of plants by an intellectual property right. By codifying intellectual property for plant breeders, UPOV aims to encourage the development of new varieties of plants for the benefit of society.

For plant breeders' rights to be granted, the new variety must meet four criteria under the rules established by UPOV:

1. The new plant must be novel, which means that it must not have been previously marketed in the country where rights are applied for.
2. The new plant must be distinct from other available varieties.
3. The plants must display homogeneity.
4. The trait or traits unique to the new variety must be stable so that the plant remains true to type after repeated cycles of propagation.

Protection can be obtained for a new plant variety (legally defined) however it has been obtained, e.g. through conventional breeding techniques or genetic engineering.

Members

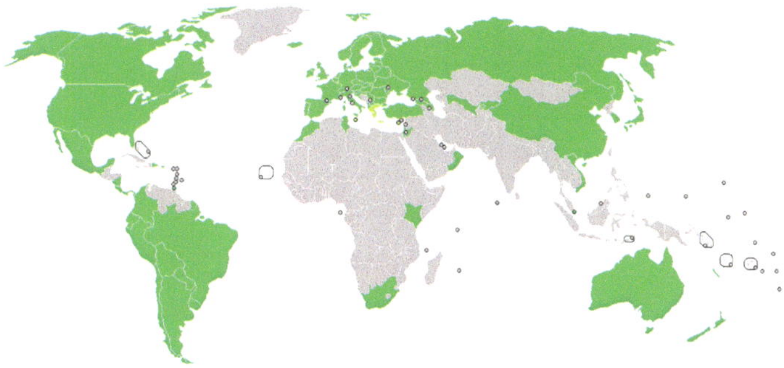

members

EU members, not separately ratified

As of October 2, 2015 the following 74 parties were members of UPOV: African Intellectual Property Organisation, Albania, Argentina, Australia, Austria, Azerbaijan, Belarus, Belgium, Bolivia, Brazil, Bulgaria, Canada, Chile, the People's Republic of China, Colombia, Costa Rica, Croatia, Czech Republic, Denmark, Dominican Republic, Ecuador, Estonia, European Union, Finland, France, Georgia, Germany, Guatemala, Hungary, Iceland, Ireland, Israel, Italy, Japan, Jordan, Kenya, Kyrgyzstan, Latvia, Lithuania, Macedonia, Mexico, Moldova, Morocco, Netherlands, New Zealand, Nicaragua, Norway, Oman, Panama, Paraguay, Peru, Poland, Portugal, Republic of Korea, Romania, Russian Federation, Serbia, Singapore, Slovakia, Slovenia, South Africa, Spain, Sweden, Switzerland, Trinidad and Tobago, Tunisia, Turkey, Ukraine, United Kingdom, United States of America (with a reservation), Uruguay, Uzbekistan, and Viet Nam.

The membership to the different versions of the convention is shown below:

Party	1961 Convention	1978 Convention	1991 Convention	Comments
Albania			15 October 2005	
Argentina		25 December 1994		
Australia		1 March 1989	20 January 2000	
Austria		14 July 1994	1 July 2004	
Azerbaijan			9 December 2004	
Belarus			5 January 2003	

Belgium	5 December 1976			
Bolivia		21 May 1999		
Brazil		23 May 1999		
Bulgaria			24 April 1998	
Canada		4 March 1991	19 July 2015	
Chile		5 January 1996		
China		23 April 1999		
China		13 September 1996		
Costa Rica			12 January 2009	
Croatia			1 September 2001	
Czech Republic		1 January 1993	24 November 2002	
Denmark	6 October 1968	8 November 1981	24 April 1998	
Dominican Republic			16 June 2007	
Ecuador		8 August 1997		
Estonia			24 September 2000	
European Union			29 July 2005	
Finland		16 April 1993	20 July 2001	
France	3 October 1971	17 March 1983	27 May 2012	
Georgia			29 November 2008	
Germany	10 August 1968	12 April 1986	25 July 1998	
Hungary		16 April 1983	1 January 2003	
Iceland			3 May 2006	
Ireland		8 November 1981	8 January 2012	
Israel	12 December 1979	12 May 1984	24 April 1998	
Italy	1 July 1977	28 May 1986		
Japan		3 September 1982	24 December 1998	
Jordan			24 October 2004	

Kenya		13 May 1999		
Kyrgyzstan			26 June 2000	
Latvia			30 August 2002	
Lithuania			10 December 2003	
Macedonia			4 May 2011	
Mexico		9 August 1997		
Moldova			28 October 1998	
Montenegro			24 September 2015	
Morocco			8 October 2006	
Netherlands	10 August 1968	2 September 1984	1 July 2004	European Netherlands only
New Zealand		8 November 1981		
Nicaragua		6 September 2001		
Norway		13 September 1993		
OAPI			24 April 1998	
Panama		23 May 1999	22 November 2012	
Paraguay		8 February 1997		
Peru			8 August 2011	
Poland		11 November 1989	15 August 2003	
Portugal		14 October 1995		
Romania			16 March 2001	
Russia.			24 April 1998	
Serbia			5 January 2013	
Singapore			30 July 2004	
Slovakia		1 January 1993	12 June 2009	
Slovenia			29 July 1999	
South Africa	6 November 1977	8 November 1981		
South Korea			7 January 2002	
Spain	18 May 1980		18 July 2007	

Sweden	17 December 1971	1 January 1983	24 April 1998	
Switzerland	10 July 1977	8 November 1981	1 September 2008	
Trinidad and Tobago		30 January 1998		
Tunisia			31 August 2003	
Turkey			18 November 2007	
Ukraine		3 November 1995	19 January 2007	
United Kingdom	10 August 1968	24 September 1983	3 January 1999	
United States		8 November 1981	22 February 1999	
Uruguay		13 November 1994		
Uzbekistan			14 November 2004	
Vietnam			24 December 2006	

System of Protection

The Convention defines both how the organization must be governed and run, and the basic concepts of plant variety protection that must be included in the domestic laws of the members of the Union. These concepts include:

- The criteria for new varieties to be protected: novelty, distinctness, uniformity, and stability.
- The process for application for a grant
- Intellectual property rights conferred to an approved breeder
- Exceptions to the rights conferred to the breeder
- Required duration of breeder's right
- Events in which a breeder's rights must be declared null and void

In order to be granted breeder's rights, the variety in question must be shown to be new. This means that the plant variety cannot have previously been available for more than one year in the applicant's country, or for more than four years in any other country or territory. The variety must also be distinct (D), that is, easily distinguishable through certain characteristics from any other known variety (protected or otherwise).

The other two criteria, uniformity (U) and stability (S), mean that individual plants of the new variety must show no more variation in the relevant characteristics than one would naturally expect, and that future generations of the variety through vari-ous propagation means must continue to show the relevant distinguishing characteristics. The UPOV offers general guidelines for DUS testing.

A breeder can apply for rights for a new variety in any union member country, and can file in as many countries as desired without waiting for a result from previous applications. Protection only applies in the country in which it was granted, so there are no reciprocal protections unless otherwise agreed by the countries in question. There is a right of priority, and the application date of the first application filed in any country is the date used in determining priority.

The rights conferred to the breeder are similar to those of copyright in the United States, in that they protect both the breeder's financial interests in the variety and his recognition for achievement and labor in the breeding process. The breeder must authorize any actions taken in propagating the new variety, including selling and marketing, importing and exporting, keeping stock of, and reproducing. This means that the breeder can, for example, require a licensing fee for any company interested in reproducing his variety for sale. The breeder also has the right to name the new variety, based on certain guidelines that prevent the name from being deliberately misleading or too similar to another variety's name.

There are explicit exceptions to the rights of the breeder, known as the "breeder's exemption clause", that make it unnecessary to receive authorization for the use of a protected variety where those rights interfere in the use of the variety for a private individual's non-monetary benefit, or the use of the variety for further research. For example, the breeder's rights do not cover the use of the variety for subsistence farming, though they do cover the use of the variety for cash crop farming. Additionally, the breeder's authorization is not required to use a protected variety for experimental purposes, or for breeding other varieties, as long as the new varieties are not "essentially derivative" of the protected variety.

The Convention specifies that the breeder's right must be granted for at least 20 years from grant date, except in the case of varieties of trees or vines, in which case the duration must be at least 25 years.

Finally, there are provisions for how to negate granted breeders' rights if the rights are determined to be unfounded. That is, if it is discovered after the application has been granted that the variety is not actually novel or distinct, or if it is discovered to not be uniform or stable, the breeder's rights are nullified. In addition, if it is discovered that the person who applied for protection of the variety is not the actual breeder, the rights are nullified unless they can be transferred to the proper person. If it is discovered after a period of protection that the variety is no longer uniform and stable, the breeder's rights are canceled.

Genetically Modified Plant Varieties

The UPOV has been updated several times to reflect changing technology and increased understanding of how plant variety intellectual property protection must work. The last revision was in 1991, and specifically mentioned genetic engineering only insofar as it is a method of creating variation. Under the UPOV Convention alone, genetically modified crops and the intellectual property rights granted to them are no different from the intellectual property rights granted for traditionally bred varieties. It is important to note that this necessarily includes the ability to use protected varieties for subsistence farming and for research.

In October 2004, two joint Symposia were held in Geneva with the World Intellectual Property Organization (WIPO). These Symposia were the WIPO-UPOV Symposium on Intellectual Property Rights in Plant Biotechnology (24 October 2003) and the WIPO-UPOV Symposium on the Co-Existence of Patents and Plant Breeders' Rights in the Promotion of Biotechnological Developments (25 October 2003). No new policy was created at either of these events, but a consensus emerged that both patents and plant-breeders' rights must combine to promote plant biotechnology.

As a policy matter, the UPOV is known to consider open and un-restricted access to the genetic resources of protected plant varieties to be important to the continued development of new varieties. This opinion is indicated in the "breeders' exemption" clause of the Convention, as described above, and was reinforced in October 2005 in a reply to a notification from the Convention on Biological Diversity.

In April 2003, the Convention on Biological Diversity asked the UPOV for comment on the use of Genetic Use Restriction Technologies (also known pejoratively as 'terminator genes') as they relate to the promotion of intellectual property rights. In the summary of their response, the UPOV stated that intellectual property protection is necessary because breeders must have the ability to recoup their money and labor investment in creating new varieties, and in that light, plants with 'terminator genes' may still be accepted for protection if they meet the other criteria. However, the UPOV comment states that the Convention and its system of protection is sufficient to protect intellectual property rights, and that with proper legal protections in place, technologies like 'terminator genes' should not be necessary.

Development and Public Interest Concerns

Whether or not UPOV negatively affects agriculture in developing countries is much debated. It is argued that UPOV's focus on patents for plant varieties hurts farmers, in that it does not allow them to use saved seed or that of protected varieties. Countries with strong farmers' rights, such as India, cannot comply to all aspects of UPOV. François Meienberg is of this opinion, and writes that the UPOV system has disadvantages, especially for developing countries, and that "at some point, protection starts to thwart development".

On the other hand, Rolf Jördens argues that plant variety protection is necessary. He believes that by joining UPOV, developing countries will have more access to new and improved varieties (better yielding, stronger resistance) instead of depending on old varieties or landraces, thus helping fight poverty and feed the growing world population.

Empirical evidence to support either point of view is lacking. However, two things are clear.

First, UPOV supports an agricultural system that is clearly export-oriented. In other word, developing countries moving towards UPOV-consistent systems tend to favour breeders who are producing for export. The example of Kenya is telling in this regard, as UPOV's own study points out, the majority of varieties are owned by foreign producers and are horticultural crops, clearly destined for export. An over-heavy dependence on agriculture for export is increasingly recognized as being unwise.

Second, given the lack of empirical evidence to support this, it would make sense to encourage debate, exchange of knowledge and research on the impacts of UPOV type plant variety protection on farming, food, human rights and other public interest objectives. However, UPOV seems to be resisting this, for instance by keeping meetings secret, not making its documents available and [refusing farmers' organisations observer status with UPOV. A recent study by Professor Graham Dutfield exploring the role of UPOV concluded that UPOV's governance falls short in many different ways. In addition, UPOV officials know very little about actual farming. They may know about breeding and favour commercial breeders, but this is not the same as knowing about how small-scale farmers actually develop new varieties and produce them. The UPOV system thus favours commercial breeders over farmers and producers, and private interests over public interests. The UN Special Rapporteur on the Right to Food, Olivier De Schutter, came to similar findings in his study of UPOV in 2009. He found that IP-related monopoly rights could cause poor farmers to become "increasingly dependent on expensive inputs" and at risk of indebtedness. Further, the system risks neglecting poor farmers' needs in favour of agribusiness needs, jeopardising traditional systems of seed-saving and exchange, and losing biodiversity to "the uniformization encouraged by the spread of commercial varieties.

References

- "About the UPOV System of Plant Variety Protection". UPOV.int. Union internationale pour la protection des obtentions végétales. Retrieved 2015-05-25.
- UPOV (19 April 2002). General introduction to the examination of distinctness, uniformity and stability and the development of harmonized descriptions of new varieties of plants (PDF) (Report). UPOV. Retrieved 29 July 2015.
- "UPOV Notification No. 69: Ratification by the United States of America of the 1991 Act". UPOV. 22 January 1999. Retrieved 5 May 2014.

- François Meienberg: Infringement of farmers' rights D+C, 2010/04, Focus, Page 156-158 Archived January 1, 2011, at the Wayback Machine.
- Rolf Jördens: Legal framework for investment D+C, 2010/04, Focus, Page 150-153 Archived January 1, 2011, at the Wayback Machine.
- UPOV to decide on farmers' and civil society participation in its sessions European Coordination Via Campesina (ECVC) & Association for Plant Breeding for the Benefit of Society (APBREBES) Archived January 10, 2011, at the Wayback Machine.
- Graham Dutfield: Food, Biological Diversity and Intellectual Property – The Role of the International Union for the Protection of New Varieties of Plants (UPOV) 2011
- NEW SECRETARY-GENERAL OUTLINES FUTURE PRIORITIES FOR UPOV, UPOV Press Release No. 77, Geneva, October 30, 2008 Archived March 26, 2010, at the Wayback Machine.

Permissions

All chapters in this book are published with permission under the Creative Commons Attribution Share Alike License or equivalent. Every chapter published in this book has been scrutinized by our experts. Their significance has been extensively debated. The topics covered herein carry significant information for a comprehensive understanding. They may even be implemented as practical applications or may be referred to as a beginning point for further studies.

We would like to thank the editorial team for lending their expertise to make the book truly unique. They have played a crucial role in the development of this book. Without their invaluable contributions this book wouldn't have been possible. They have made vital efforts to compile up to date information on the varied aspects of this subject to make this book a valuable addition to the collection of many professionals and students.

This book was conceptualized with the vision of imparting up-to-date and integrated information in this field. To ensure the same, a matchless editorial board was set up. Every individual on the board went through rigorous rounds of assessment to prove their worth. After which they invested a large part of their time researching and compiling the most relevant data for our readers.

The editorial board has been involved in producing this book since its inception. They have spent rigorous hours researching and exploring the diverse topics which have resulted in the successful publishing of this book. They have passed on their knowledge of decades through this book. To expedite this challenging task, the publisher supported the team at every step. A small team of assistant editors was also appointed to further simplify the editing procedure and attain best results for the readers.

Apart from the editorial board, the designing team has also invested a significant amount of their time in understanding the subject and creating the most relevant covers. They scrutinized every image to scout for the most suitable representation of the subject and create an appropriate cover for the book.

The publishing team has been an ardent support to the editorial, designing and production team. Their endless efforts to recruit the best for this project, has resulted in the accomplishment of this book. They are a veteran in the field of academics and their pool of knowledge is as vast as their experience in printing. Their expertise and guidance has proved useful at every step. Their uncompromising quality standards have made this book an exceptional effort. Their encouragement from time to time has been an inspiration for everyone.

The publisher and the editorial board hope that this book will prove to be a valuable piece of knowledge for students, practitioners and scholars across the globe.

Index